MIX
Papier aus verantwortungsvollen Quellen
Paper from responsible sources
FSC® C105338

Rolf Kretschmann
Daniel Wrobel

Attitudes and Perceptions toward Physical Education

A Study in Secondary School Students

Anchor Academic
Publishing

Kretschmann, Rolf, Wrobel, Daniel: Attitudes and Perceptions toward Physical Education: A Study in Secondary School Students, Hamburg, Anchor Academic Publishing 2015

Buch-ISBN: 978-3-95489-465-9
PDF-eBook-ISBN: 978-3-95489-965-4
Druck/Herstellung: Anchor Academic Publishing, Hamburg, 2015

Cover picture: www.pixabay.com
Cover design: Anna Klöhn

Bibliografische Information der Deutschen Nationalbibliothek:
Die Deutsche Nationalbibliothek verzeichnet diese Publikation in der Deutschen Nationalbibliografie; detaillierte bibliografische Daten sind im Internet über http://dnb.d-nb.de abrufbar.

Bibliographical Information of the German National Library:
The German National Library lists this publication in the German National Bibliography. Detailed bibliographic data can be found at: http://dnb.d-nb.de

All rights reserved. This publication may not be reproduced, stored in a retrieval system or transmitted, in any form or by any means, electronic, mechanical, photocopying, recording or otherwise, without the prior permission of the publishers.

Das Werk einschließlich aller seiner Teile ist urheberrechtlich geschützt. Jede Verwertung außerhalb der Grenzen des Urheberrechtsgesetzes ist ohne Zustimmung des Verlages unzulässig und strafbar. Dies gilt insbesondere für Vervielfältigungen, Übersetzungen, Mikroverfilmungen und die Einspeicherung und Bearbeitung in elektronischen Systemen.

Die Wiedergabe von Gebrauchsnamen, Handelsnamen, Warenbezeichnungen usw. in diesem Werk berechtigt auch ohne besondere Kennzeichnung nicht zu der Annahme, dass solche Namen im Sinne der Warenzeichen- und Markenschutz-Gesetzgebung als frei zu betrachten wären und daher von jedermann benutzt werden dürften.

Die Informationen in diesem Werk wurden mit Sorgfalt erarbeitet. Dennoch können Fehler nicht vollständig ausgeschlossen werden und die Diplomica Verlag GmbH, die Autoren oder Übersetzer übernehmen keine juristische Verantwortung oder irgendeine Haftung für evtl. verbliebene fehlerhafte Angaben und deren Folgen.

Alle Rechte vorbehalten

© Anchor Academic Publishing, Imprint der Diplomica Verlag GmbH
Hermannstal 119k, 22119 Hamburg
http://www.diplomica-verlag.de, Hamburg 2015
Printed in Germany

List of Contents

List of Figures 3

List of Tables 5

Abstract 8

1 Introduction 9

2 Review of Current Literature 13

2.1 Definition of Key Terms and Concepts 13
2.1.1 Attitude 13
2.1.2 Perception 18
2.1.3 Elements of Physical Education Classes 22
2.1.4 Physical Activity and Exercise 28

2.2 Importance of Physical Activity and Physical Education 30
2.2.1 Health Behavior and Physical Education 31
2.2.2 Students' Self-Perception in Physical Education 34

2.3 Current Research on Physical Education Classes 36
2.3.1 Motivation and Physical Education 37
2.3.2 Research on German Physical Education Classes 39
2.3.3 Students' Attitudes toward Physical Education 45

2.4 Research Instruments to Measure Attitudes 54
2.4.1 Quantitative Instruments for Assessing Attitudes toward Physical Education .. 55
2.4.2 Qualitative Instruments for Assessing Attitudes toward Physical Education 59

3 Study Design 61

3.1 Questions and Hypotheses 61

3.2 Method Outline 67
3.2.1 Structure of the Questionnaire 67

3.2.2	Pretest	71
3.2.3	Procedure of the Study	73
3.2.4	Interpretation of Variables	75
3.2.5	Sample	78
3.3	Statistics	82
4	Results ..	83
4.1	Descriptive Statistics	83
4.2	Inferential Statistics	94
5	Discussion	117
6	Conclusion	127
	References	130
	Appendices	145
	Author Information	145

List of Figures

Figure 1: A scheme of the Theory of Reasoned Action (A) and the Theory of Planned Behavior (B) (Source: Madden et al., 1992, p.4). 15

Figure 2: The Perception Processing System (Source: Pickens, 2005, p. 57). 20

Figure 3: The Mutual Influence of Attitude and Perception (Source: Rokeach, 1973, as printed in Seel, 2003, p. 125). 21

Figure 4: Differentiation between Physical Activity and Exercise (Source: Caspersen et al., 1985, p. 127). 30

Figure 6: The Physical Self-Perception and its subdomains (Source: Fox & Corbin, 1989, as printed in Raustorp et al., 2005, p. 127). 35

Figure 7: The different Types of Motivation according to the Self-Determination Theory (Source: Deci & Ryan, 1985, as printed in Vallerand & Losier, 1999, p. 153). 39

Figure 8: Sporting activity of adolescents', separated according to gender (Source: Lampert et al., 2007, p. 639). 41

Figure 10: Results of the Pretest, displaying Complaints of the Pretest Group and corresponding Numbers of Choices, for the Whole Group and Separated according to Gender. 72

Figure 11: A classification of the BMI (Source: World Health Organization, 2000, p. 9). 76

Figure 12: Sample Composition regarding the Type of School. 79

Figure 11: Frequency Distribution of Students' Age. 80

Figure 12: Distinction of the Sample with reference to citizenship. 81

Figure 15: Distribution of Students' BMI. 84

Figure 14: Distribution of Students' SES. 85

Figure 17: Distribution of Students' Grade Point Averages. 86

Figure 18: Distribution of Students' Attitudes. 93

Figure 19: Simple Linear Regression for the variables Attitude and Physical Education Grade. 100

Figure 20: Attitudes Separated according to Migration Background. 102

Figure 21: Mean Attitudes for the two SES groups. 104

Figure 22: Mean Attitudes for the Three Classified BMI Groups. 106

Figure 23: Mean Attitudes Separated according to Exercise Frequency. 108

Figure 22: Mean Attitudes Separated according to Exercise Frequency. 110

Figure 25: Mean Attitudes Separated according to Students' Parents' Physical Activity Frequency. 113

List of Tables

Table 1: Overview of Studies on Students' Attitudes toward Physical Education. 50

Table 2: Instrument for assessing students' attitudes towards physical education (Subramaniam & Silverman, 2000) 58

Table 3: Gender Distribution of the Sample 78

Table 4: Sample Composition regarding the Type of School. 79

Table 5: Descriptive Statistics of the Students' Age. 80

Table 6: Students' BMI as classified according to the WHO (2000, 9). 85

Table 7: Distribution of Students' Physical Education Grades. 87

Table 8: Exercise Behavior of Students. 88

Table 9: Physical Activity Behavior of Students 88

Table 10: Exercise Behavior of Students' Peers. 89

Table 11: Physical Activity Behavior of Students' Peers. 89

Table 12: Exercise Frequency of Students' Parents. 90

Table 13: Physical Activity Frequency of Students' Parents 90

Table 14: Statistics of the Different Conceptualizations of Attitude. 92

Table 15: Results of the t-test for Random Samples. 95

Table 16: t-test investigating the differences regarding gender. 96

Table 17: Descriptive Statistics regarding the Different Types of School. 96

Table 18: One-way ANOVA examining differences regarding the types of school. 97

Table 19: Post-hoc Test regarding the Types of School. 97

Table 20: Spearman's Rho and Significance for Grade Point Average and Attitude. 99

Table 21: Spearman's Rho and Significance for Physical Education Grade and Attitude. 100

Table 22: t-test Investigation of the Differences in Attitude regarding Migration Background. 102

Table 23: t-test Investigation of the Differences in Attitude regarding the SES. 104

Table 24: t-test Investigation of the Differences in Attitude regarding the Classified BMI. 106

Table 25: Data of the post-hoc Test for Differences concerning Students' Exercise Behavior. 109

Table 26: Data of the post-hoc Test for Differences concerning Students' Exercise Behavior. 109

Table 27: Data of the post-hoc Test for Differences concerning Students' Parents' Exercise Behavior. 112

Table 28: Descriptive Data concerning Students' Attitudes divided according to Students' Peers' Exercise Behavior. 114

Table 29: Data of the Post-Hoc Test for Differences concerning Students' Peers' Exercise Behavior. 115

Table 30: Descriptive Data concerning Students' Peers' Physical Activity Behavior... 115

Table 31: Data of the post-hoc Test for Differences concerning Students' Peers' Physical Activity Behavior. 116

Abstract

Physical education teaching and learning efforts obviously target the student. Like parents, teachers, administrators and any other directly or indirectly involved parties, students do have opinions based on their experience on their respective physical education classes and physical education in general. These opinions, or so-called attitudes, are important to research due to their potential of giving insight to the learner's perspective, which may also serve as an authentic feedback from the student. This study investigated German secondary school students' attitudes toward physical education. Results have the intention to reveal what attitudes towards physical education German students have and which factors influence these attitudes. The study sample contained students from the different school types Gymnasium, Realschule, and Haupt-/Werkrealschule. The students were surveyed via questionnaire that was developed based on validated research instruments from prior studies in the field. Data was analyzed integrating independent variables such as students' gender, physical education grade, grade point average, body mass index, socioeconomic status, type of school, citizenship, and the exercise and physical activity behavior of students, their parents and their peers.

Keywords: physical education, attitude, school students, student attitudes, secondary school students, attitude measures

1 Introduction

Investigating physical education means to take all involved parties and individuals into account. Physical education teachers and their students will usually come into one's mind when thinking of direct involvement in regular physical education classes. Parents, principals, administrators, college level lecturers and professors, and policy makers may also be considered for mutual interactions and discourse regarding physical education.

Although students are certainly the primary recipients of physical education teaching efforts, only few research-oriented efforts have been attempted to tackle the students' perspectives regarding physical education on the contrary (Dyson, 2006). Despite the rich empirical research tradition on physical education, there is only little research on the point of view of the students, and their perceptions and attitudes towards physical education (Graber, 2001).

In 2006, the DSB published the *SPRINT-Studie* as one of the most recent and most frequently discussed German empirical investigation of physical education classes.[1] The study examined several aspects of physical education classes such as curriculum, state of sports facilities and others (cf. Brettschneider & Kuhlmann, 2006, 12ff). The point of view of the students and their perceptions and attitudes towards physical education, however, was severely neglected. Besides the *SPRINT-Studie* (Deutscher Sportbund, 2006), almost no studies in the field of physical education classes exist in Germany. In

---

[1] SPRINT is the abbreviation for **Sp**ortunterricht **in** Deutschland and was commissioned by the DSB as a result of a lack of empirical investigation in the field of physical education classes. The research was conducted by eight research groups from different German universities. Purpose of the study was to investigate the situation of physical education classes in German schools and to offer a variety of guidance how to improve the current situation (cf. DSB, 2006). See also chapter 2 for further discussion.

contrast, on an international level, many investigations have been conducted, all of which focus on different aspects of physical education.

Those studies, which are concerned with the perceptions and attitudes of students towards physical education classes basically solely focus on the methods to examine perceptions and attitudes and the general shape of perceptions and attitudes (cf. e.g., Phillips & Silverman, 2012; Subramaniam & Silverman, 2000).

The importance of students' positive attitudes towards physical education emerges out of the suggestion that students who developed a positive attitude to physical education and physical activity will more likely adopt a physically active lifestyle (Solmon & Lee, 1996; Wallhead & Buckworth, 2004). In addition, a positive attitude to one subject, in this case physical education may also indirectly influence students' attitudes towards school and education in general (Bibik, Goodwin, & Orsega-Smith, 2007), as well as academic achievement (Howie & Pate, 2012).

The influence of different factors on the perceptions and attitudes of students was hardly taken into account. Since other studies in the field of physical education have shown that e.g., gender has a major influence on self-perception or motivation[2], it seems to be logical to conclude that several factors do influence the perceptions and attitudes of students, as well. The important question that needs to be examined therefore is, which factors influence the students' perceptions and attitudes towards physical education classes.

Starting from this question, several other important questions arise: how can perceptions and attitudes be measured? What are attitudes and perceptions? How can we define perceptions and attitudes? Are those perceptions and attitudes also influenced by

---

[2] International studies in the field of physical education classes have shown that e.g., the self-perception of students is influenced by gender and age (cf. e.g., Maiano, Ninot, & Bilard, 2004). See also chapter 2 for further discussion.

the degree of physical and sporting activity? Are they influenced by the parents' or the friends' point of view.

These questions are especially interesting for teachers and future teachers. They should know how attitudes are formed and how they are developed in reality. This knowledge should help to understand complex contexts of physical education classes, as well as the varying behavior of students. Besides, it would be helpful to know how the different aspects of physical education classes influence the perceptions and attitudes of students. Content of the curriculum or the behavior of the teacher may be such aspects, only to name a few. The present study's underlying idea of perceptions and attitudes consisting of two components[3] could reveal, which aspects students place value on and by which aspects they are influenced emotionally. The answers to the above questions should also make teachers more sensitive to the reasons of students' behavior.

The here presented study is interested and concerned with those questions, as well. Its purpose is not only to find out how perceptions and attitudes of students are shaped, but to a greater degree to reveal which factors influence the perceptions and attitudes of students. The present study therefore picks up the above raised questions to further investigate the field of physical education classes. In the first part, the theoretical foundation for the study will be presented. On the basis of current literature, the current status of research in the field of physical education will be reviewed to lay ground for the empirical part of the study. First, key terms and concepts that are important for the investigation of the perceptions and attitudes will be defined to mark the boundaries of the research. Since the purpose of this study is to provide results not only for the German research field, but on an international level, the subject of German physical education is defined thoroughly. The paper will then turn to the current findings related to motivation, self-perception, health, and eventually to the perceptions and attitudes in

---

[3] Different models have been proposed for the definition of attitudes (cf. e.g., Oppenheim, 2000). The different approaches and the here-applied model will be further discussed in chapter 2.

the context of physical education classes. It is then discussed, how methods and instruments can be used to assess and measure attitudes and perceptions. In the second part of the paper, the actual empirical study will be presented. It contains the above raised research questions and underlying hypothesis and describes the procedure of the study. Therefore, the methods and instruments that have been used to conduct the present study are expounded. Afterward, the results of the study will be described and analyzed, before, in the last part, the results will be discussed regarding the purpose of the study and in the context of existing literature and already conducted studies.

2 Review of Current Literature

In this chapter, the theoretical basis for further investigation and the empirical study are lined out. All the key terms and concepts that are essential to understand and work with the subject of the present study are defined in this chapter. It will then turn to the current state of research and discuss the importance of motivation, self-concept and health behavior for physical activity and exercise, and for physical education. Afterward, findings on perceptions and attitudes toward physical education classes will be lined out. Furthermore, it will also look at the possibilities of investigation and at the – so far – proposed and used methods and instruments.

2.1 Definition of Key Terms and Concepts

It might look easy and trivial to define the terms 'perception' or 'attitude', but the further process of investigation is in fact determined by how those terms are defined. Besides, the question of how German physical education classes are determined, or to be more precise, of what they consist, influences the empirical study. Moreover, a distinction between physical activity and exercise is necessary to understand existing research and the instrument used in the present study. As already outlined above, this section will deal with those key terms and concepts that need to be taken into account to further investigate the whole field of perceptions and attitudes of students toward physical education classes.

2.1.1 Attitude

The term 'attitude(s)' is omnipresent in our everyday lives, from a psychological point of view, however, the term has not been used correctly in society (cf. Silvermann & Subramaniam 1999, p. 97). In fact, attitude has been one of the most important subjects in social psychology (cf. Maderthaner, 2008, p. 337), and even within the field of science, attitude has always been a concept of much controversy.

From a psychological point of view, however, the term has not been used correctly in society (Silverman & Subramaniam, 1999). In fact, attitude has been one of the most important subjects in social psychology (Allport, 1935; Myers, 2012), and even within the field of science, attitude has always been a concept of much controversy. As Fishbein and Ajzen (1975, p. 1) have stated, it "is characterized by an embarrassing degree of ambiguity and confusion". As the history of attitude related research indicates, attitudes have a major influence on behavior (Ajzen, 2007), so it is crucial to understand and illustrate the role attitudes occupy in shaping and affecting behavior. The significance that is attributed to attitude relating to behavior shaping, however, differs between the various psychological approaches.

The *Theory of Reasoned Action*, which can be traced back to the works of Martin Fishbein with the assistance of Icek Ajzen (1975), has been widely used and accepted to describe the relation between attitude and behavior. In a nutshell, this theory assumes that behavior can be described as a reasoned process, in which a person forms two kinds of beliefs, that is behavioral and normative beliefs. The prior are beliefs about the consequences of behavior, the latter are beliefs about important persons' thoughts relating to the realization of the intended behavior (cf. Trafimov, 2007, p. 24). Trafimov (2007, p. 24) further states that people also evaluate the consequences of behavioral beliefs and the importance of conforming to other people. The sum of all those normative beliefs and their evaluations is called the subjective norm and the sum of all behavioral beliefs and the corresponding evaluations is called attitude. Together, those two factors determine the behavioral intention, which in turn influences the realization of a particular behavior (cf. Madden, Ellen, & Ajzen, 1992, p. 3).

Figure 1 shows models of the *Theory of Reasoned Action* and of the *Theory of Planned Behavior*.

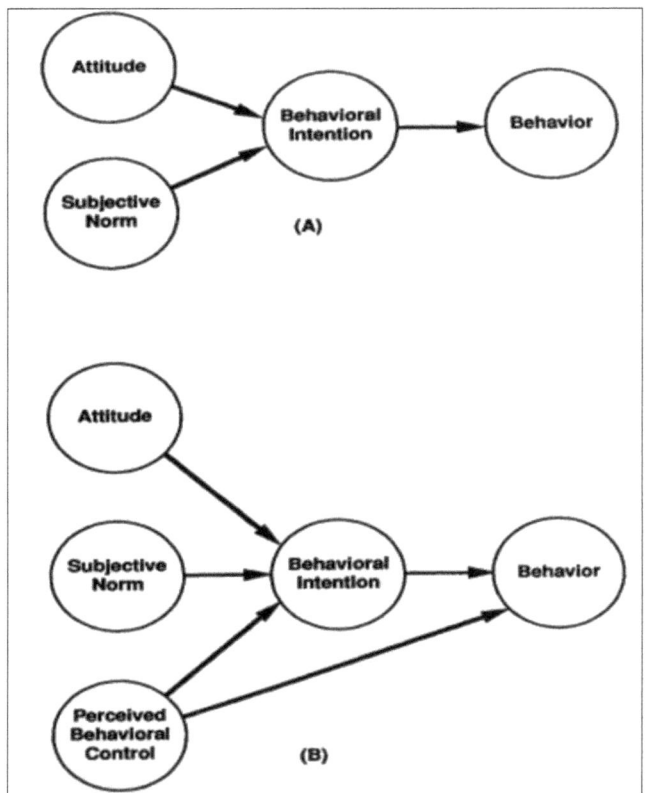

Figure 1: A scheme of the Theory of Reasoned Action (A) and the Theory of Planned Behavior (B) (Source: Madden et al., 1992, p.4).

Restrictions to the *Theory of Reasoned Action* have been posed, however, such that this theory applies only to voluntary acts (Sheppard, Hartwick, & Warshaw, 1988). Ajzen (1985) proposed an extension of this model, including perceived behavioral control into the scheme of the formation of behavior. In short, the introduction of this additional module extends the above expressed border of purely intentional acts (Madden et al.,

1992). The important conclusion of this theory is that attitude is in fact a major determinant of behavior. The knowledge about the attitude of a person is therefore important to adapt oneself to the behavior of others (cf. Leiße, 2006, p. 159; Smith & Mackie, 2007).

Other approaches deal with the conceptualization of the isolated module of attitude to come to terms with what attitude itself actually displays. According to Eagly and Chaiken (1998), attitude is the summarized evaluation of an object. It is an individual's learned tendency to evaluate a phenomenon a certain way. Basis for an attitude is the existence of a particular object towards which an attitude is formed and the process of an evaluation. The prior can represent any matter people can imagine or perceive, abstract or concrete. From this chapter's perspective, this reference object is the physical education class. On the other hand, the evaluation of the object at hand is the decisive connection between the latter and the resulting behavior (Fazio & Towles-Schwen, 1999). To explain the process of this evaluation and thus the formation of attitudes, different models have been introduced.

The proponents of the single-component model see attitude as uni-dimensional construct (Bohner & Wänke, 2002). The only component taken into account in this approach is usually the affective dimension (Subramaniam & Silverman, 2000). The affective dimension displays feelings and emotions and the corresponding evaluation of objects. Fazio and Zanna (1981, p. 162) define attitude as "evaluative feeling that is evoked by a given object" which shows the above mentioned reference to emotions. Furthermore, most of the research in the field of attitude relied on this uni-dimensional concept of attitude (Bagozzi & Burnkrant, 1979).

Other researchers suggested a two-component view of attitude (Oppenheim, 2000). In addition to the affective dimension, a second, so called cognitive dimension is introduced. Whereas the affective dimension again refers to the emotions and feelings a person has towards an object, the cognitive dimension displays beliefs and knowledge about the characteristics of the object at hand (Hogg & Vaughan, 2008, p. 149).

According to Bohner (2002, p. 267), attitude is – as mentioned above – "the summarized evaluation of an object".[4] Basis for an attitude is the existence of a particular object toward which an attitude is formed and the process of an evaluation. The prior can represent any matter people can imagine or perceive, abstract or concrete (cf. Bohner, 2002, p. 267). In the present study, this reference object is the physical education class. On the other hand, the evaluation of the object at hand is the decisive connection between the latter and the resulting behavior (cf. Meinefeld, 1977, p. 17). To explain the process of this evaluation and thus the formation of attitudes, different models have been introduced.

The proponents of the single-component model see attitude as uni-dimensional (cf. e.g., Bohner & Wänke, 2002). The only component taken into account in this approach is usually the affective dimension (cf. Subramaniam & Silvermann, 2000, p. 30). The affective dimension displays feelings and emotions and the corresponding evaluation of objects. Fazio and Zanna (1981, p. 162) define attitude as "evaluative feeling that is evoked by a given object", which shows the above mentioned reference to emotions. Meinefeld (1977, p. 37) further indicates that most of the research in the field of attitude relies on this uni-dimensional concept of attitude[5], which is confirmed by Bagozzi and Burnkrant (1979).

The third approach explaining attitude is the multi-component model (cf. e.g., Triandis, 1971). Besides the above-mentioned affective and cognitive dimension, a behavioral

---

[4] Translated from German; the original passage quotes as follows: "[...] definieren Einstellung als eine zusammenfassende Bewertung eines Gegenstandes [...]"(Bohner, 2002, p. 267).

[5] Moreover, Meinefeld (1977, p. 37) also traces the uni-dimensional concept of attitude and the mentioned uni-dimensional measurement back to one of the most influential psychologists, Louis Thurstone, who defined attitude as an emotion related to a particular object.

component exists. This behavioral component comprises both actions toward the object and intentions to act (Ajzen, 2007, p. 4).[6] However, this model has been criticized by various researchers as placing too much emphasis on the overt behavior, whereas in empirical studies the highest correlations were usually found between the attitude and the behavioral intention (McGuire, 1989, p. 41).

Despite these various approaches, however, it is important to emphasize the role attitude plays in behavior formation. The importance of this role might vary between the approaches, but no matter which model is used to describe attitude, the concept is always given an influencing role concerning behavior formation. In this study, the two-dimensional concept of attitude was used to investigate students' attitudes toward physical education classes.[7]

2.1.2 Perception

Similar to the above-discussed attitudes, the usage of the term perception is common in ordinary speech. Again, the question is how to define this concept to take hold of a universal understanding of this concept. As Hagendorf, Müller, Krummenacher, and Schubert (2011) suggest, a perception is not just an accurate reflection of the external world, but rather a complex procedure.[8] Perception can be defined as "a process of interpreting and organizing information provided by the sensory organs" (Hagendorf et

---

[6] The three dimensions of attitude in this model are also labeled as response categories. It is worth noting at this point that within this multi-dimensional model psychologists also distinguish between a verbal and a non-verbal response mode. The fact that it is typical to use cognitive or affective responses on the verbal side and behavioral responses on the non-verbal side accounts for the difference of what people say and how they actually act. Whereas critics have often argued that the verbal mode reflects the attitude and the non-verbal mode reflects behavior, advocates of this model have claimed both to be observable behaviors, and that both mirror an adjacent attitude (cf. Ajzen 1989, 244f).

[7] Cf. chapter 2.3 for more precise discussion.

[8] Cf. Sternberg & Sternberg (2012).

al., 2011, p. 5).⁹ Perception is usually seen as an unconscious process, even if we – of course – sometimes perceive our environment consciously. As the quotation above indicates, the information that should be processed must necessarily be provided by involved organs. The sensory organs and the brain register the perceivable (physical and chemical) stimuli (cf. Scharfetter, 2010, p. 184).

On a physiological level, the processing of information from sensory organs is regulated by neurons (cf. Pomerantz, 2003; Saladin 2007, p. 443).¹⁰ At the same time, the perception does not, as indicated above, reflect all the stimuli around us, but selects those information that seem to be important in the respective situation (cf. Hagendorf et al., 2011, p. 8; Kolb & Whishaw, 2009). To determine the importance of perceivable information, the latter is mixed with feelings and experiences (cf. Hoffmann 2004, p. 18; Sternberg & Sternberg, 2012).

In brief, a perception basically is the absorption of an external stimulus, the recognition of this stimulus by the brain according to previous experiences, its organization, and interpretation (cf. Pickens, 2005, p. 52). Result of this procedure is that various participants have different perceptions of the same situation, and that the perception always refers a purely subjective concept.

Figure 2 displays the complex process of the perception processing system.

---

[9] Translated from German; the original passage quotes as follows: "Wahrnehmung ist ein Prozess, mit dem wir die Informationen, die von den Sinnessystemen bereitgestellt werden, organisieren und interpretieren." (Hagendorf et al., 2011, p. 5).

[10] Of course, this description only superficially touches the topic of the complex nervous system. At this point, however, it does not seem necessary to go into detail concerning this topic. For further information see Saladin (2007). Similarly, the various physical and chemical theories of perception are not mentioned here. See Robinson (1994) for further information concerning this topic.

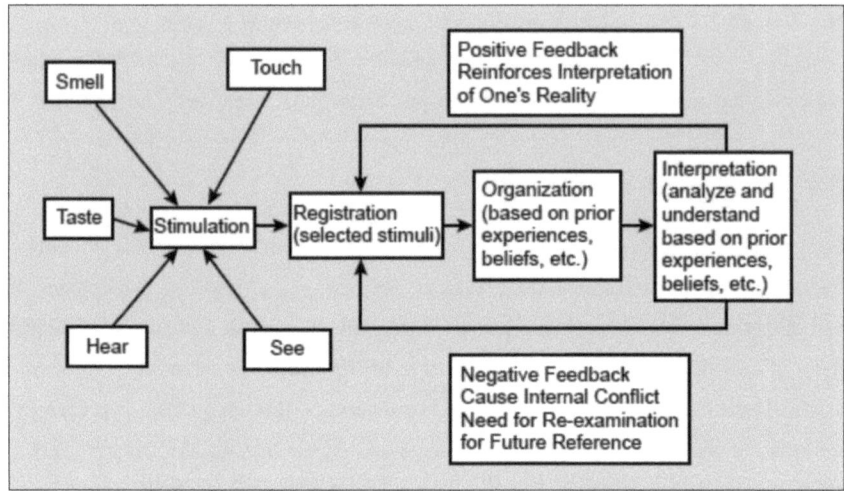

Figure 2: The Perception Processing System (Source: Pickens, 2005, p. 57).

As the above connection already indicates, perceptions seem to be influenced by a variety of other concepts. Hausmann (2009, p. 27) has listed different factors influencing perceptions to a certain degree.[11] The more knowledge one has about a certain object or situation, the more one will perceive of this object. If we expect certain things to happen, we will already perceive the initial indicators of the corresponding information. Needs and wishes do also reinforce the perception of related stimuli. Last but not least, the attitude filters what we perceive. People will above all perceive that information that accord with their attitudes. One of the best examples how attitudes influence perception are prejudgments and stereotypes. Both are generalizing attitudes that result in corresponding emotions and behavior, and at the same time influence what

---

[11] Cf. Johns & Saks (2008).

one perceives of the respective object (cf. Hockenbury & Hockenbury, 2007; Maderthaner 2008, p. 338).

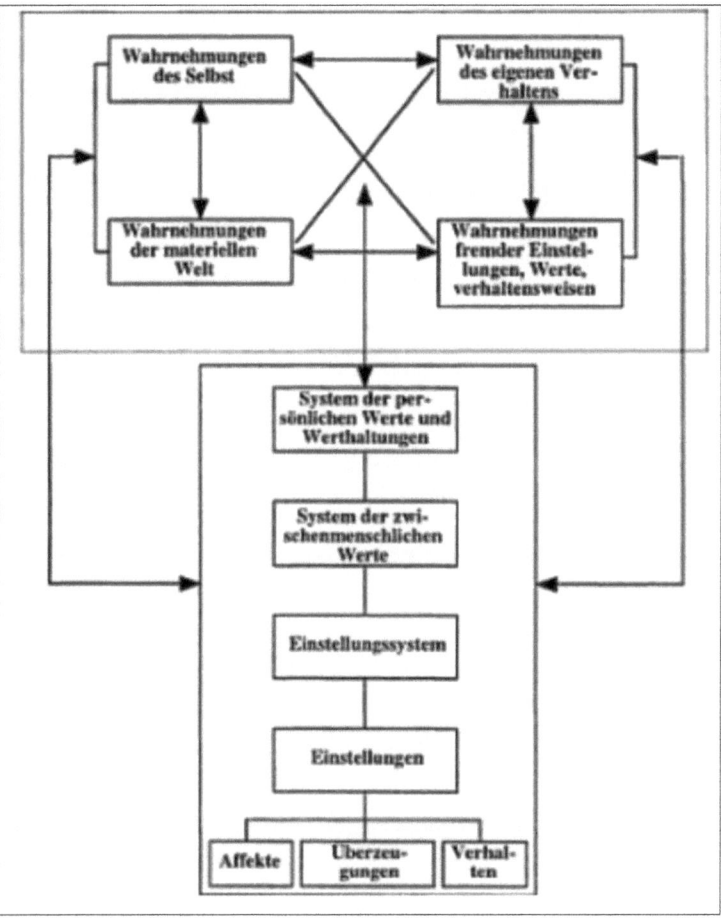

Figure 3: The Mutual Influence of Attitude and Perception (Source: Rokeach, 1973, as printed in Seel, 2003, p. 125).

At the same time, Seel (2003, p. 125) indicates that perception also influences attitude.[12] Since attitudes need to be formed sometime, they are necessarily preceded by the perception of information concerning the object the attitude is formed on. These perceptions include perceptions of oneself, the own behavior, the external world and attitudes and behaviors of others (cf. Seel, 2003, p. 125). Table 3 shows the mutual influence of attitude and perception (Rokeach, 1973, as printed in Seel, 2003, p. 125).

As we have seen, perception is a difficult concept to grasp. When there is talk of perceptions in this study, it is important to relate this to the mentioned subjective concept of absorbing selected stimuli and information depending on one's previous knowledge, experiences and feelings. One of the most important facts is its relation to attitude. Proof has been given – as illustrated above – that the two concepts are interdependent, that is attitude influences perception and by the same token is influenced by the latter (Bohner & Wänke, 2002).

2.1.3 Elements of Physical Education Classes

School sport and physical education classes constitute a broad field, in which a lot of research has been conducted and many theories have been established. The first condition that needs to be imposed on the subject here is the restriction of the subsequently discussed range of physical education classes to Germany. Since a German University realized this study and the data conducted in German schools, this focus on a definition of German physical education classes is essential.

Among the most intensively investigated areas are the legitimation and the function of physical education classes (cf. e.g., Balz & Neumann, 1999), concepts and methods of

---

[12] Cf. Pickens (2005).

school sports (cf. e.g., Neumann & Balz, 2004) or the elements of good physical education classes (cf. Gebken, 2005).[13]

The question, which is relevant for this study, though, is the determination of central modules that constitute the every day physical education class. To understand the perceptions and attitudes of students toward the physical education class, the answer to the question at hand is necessary. Only then is it possible to correctly investigate the attitudes of students toward the intended object – that is, physical education classes in Germany.

Scherler (2000) presents eight key elements of the institutionalized sport and thus tries to define what physical education actually means. Thienes (2008) takes this up and further explains those key elements.

The first element is its obligatory status. Students are obliged to participate in physical education classes – to be more precise not only to attend the classes but also to participate actively. The amount of physical education classes per week varies from state to state, from school to school and even within the different grades.[14]

The second feature of school sport is the goal orientation. Rather than being executed for a self-purpose, physical activity in school is connected to specialist purposes. This feature is of course closely related to the above-mentioned legitimation of physical

---

[13] Event though the above-mentioned fields of research basically do not play a decisive role in investigation of students' attitudes, it is worth mentioning the central ideas of a few. Balz and Neumann (1999) describe a dual function of physical education. See also below in this chapter for further explanation. Meanwhile, Gebken (2005) identifies 8 elements of good physical education classes: a clear structure; an optimal time of exercise for students; a variety of methods; coherence of goals, contents and methods; good atmosphere; the usage of feedback; individual support; transparency of expectations and control of performances.

[14] Interestingly enough, the amount of physical activity outside school may vary from student to student. This unequal distribution and thus the unequal competence of various students leads to one of the school sport's most important dilemmas, that is its acceptance. See also chapter 2.2 for further information.

education classes. It is those specialist purposes that legitimate school sport because it serves a higher purpose, on a subject-specific or a scholastic level.[15]

The third issue is the selection of the content of physical education classes. This content is partly regulated by curricula of the concerned state, however, the individual schools can define focuses for their physical education classes within a given frame.

Fourth, the need for marks – as in all the other school subjects – is a major constituent of physical education class. In Germany, however, the mark for school sports alone is not relevant for moving up in the next grade. Since the grading in general is a controversial issue, much research has been conducted in the field of physical education class grading and many recommendations have been made (cf. e.g., Balz & Kuhlmann, 2003).

The fifth property is the employment of specialist subject teachers and thus the management by an expert. Physical education teachers have to adopt several roles at once, and, as the facilitator, much depends on the ability of the teacher. According to Bräutigam (2011a, p. 21), the main tasks of a physical education teacher are teaching, educating, assessing, advising and innovating. As a result, the teacher is responsible for the realization of the above-mentioned elements and the legitimation of physical education classes. Moreover, as a matter of fact, teachers are usually subject to evaluation and criticism of students, parents and other observers (cf. Balz, Bräutigam, Miethling & Wolters, 2011, p. 121).

The grouping in age groups constitutes the sixth element of the physical education class (cf. Thienes, 2008, p. 236). This aspect, though, is of course due to the organization of the school as institution. Concerning school sport, this element may be further restricted to groupings according to gender. Most schools have – at least from grade 7 to grade 11

---

[15] A closer examination of the legitimation of school sports follows later in this chapter.

– grades split up in male and female physical education grades. This merging of all the grades of one level leads to the students getting in touch with peers of parallel classes, as well. Thus, students do not only come in contact with their well-known classmates, but also with other students from other classes. Besides the sport's well-known quality of social interaction, physical education classes can therefore be seen as a field of even more and enhanced interaction.

The last two characteristics Scherler (2000) presents are rather limitations to the physical education class, namely the time restriction and the spatial and material situation. The time restriction results from the nature of the subject: getting to the sport facilities and getting dressed for the class reduces the time left for physical education drastically. In a 45-minute class, the above-mentioned restrictions shorten the time left to 35 minutes (cf. Scherler, 2000, p. 53). Taking into account that two to three lessons per week are the average amount of physical education classes, this fact gives even more importance to the time situation of school sport, especially with regard to the fact that many topics cannot be treated in 45 minutes (cf. Thienes, 2008, p. 240).

The spatial and material problem results from the fact that physical education classes can neither be performed in the normal class room nor – due to legal and organizational issues – in special public places. Thus, the physical education classes have to take place in gyms – or, in summer, on other sports fields. The material conditions of these facilities, in turn, influence the content of school sport. Findings of the SPRINT-Studie (DSB, 2006, p. 14) moan a severe restriction of school sport due to the spatial and material situation.

These constitutive elements of physical education classes are necessary to understand what physical education classes actually embody, and they are therefore the key to assess and investigate students' attitudes and perceptions toward the subject of school sport.

In addition to the above discussed, the principles and the function of the institutionalized physical and sporting activity is an essential element of the academic didactic discussion and moreover of this study. As already indicated in this paper, the legitimation of school sport follows from the fact that this subject is not included in the curricula for a self-purpose, but for special purposes.[16] This multi-perspective view of physical education is derived from the dual function of school sport. On the one hand, students are to be educated to become sympathetic with sporting activity, on the other hand they are to be educated by means of physical activity (cf. Balz & Neumann, 1999).[17] Physical activity is a broad field and its participants can gain important experiences from engaging in this activity. Thus, "the principle of multi-perspectivity results from the fact that sporting activity is a multifaceted field of activity which addresses all the senses" (Bräutigam, 2011a, p. 86).[18]

In other words and related to the physical education class, this means that students should obtain the possibility to experience the whole field of sporting activity in various ways. Kurz (2004, pp. 66f) formulates six perspectives that should be addressed during physical education classes:

- experience and reflect performance and achievement

- to physically express oneself and to shape movement

- to dare something and to take responsibility for action

---

[16] As described above, the review of literature referring to the questions at hand is limited to German literature. Depending on the country, other authors stress different purposes of physical education classes. Simons-Morton (1994), for example, highlights the participation of students in physical activity and the acquisition of skills to live an active lifestyle as the major aims of school sport.

[17] This dual function is in German referred to as "Erziehung zum Sport und Erziehung durch Sport" (cf. Balz & Neumann, 1999).

[18] Translated from German; the original passage quotes as follows: "Das Prinzip der Mehrperspektivität ergibt sich aus dem Umstand, dass Sport ein vielseitiges und vielsinniges Tätigkeitsfeld ist." (Bräutigam, 2011a, p. 86).

- to improve the ability of perception and to extend the experience of movement

- to enhance the personal level of fitness and to develop a sense for the effects on health

- to experience cooperation, communication and competition

Since the pedagogic perspectives are not organized hierarchically, but rather display an equal system, no perspective is to be privileged. Some contents of the physical education class do, of course, permit a more natural relation to certain perspectives, such as track and field with its cgs-system[19] of units and the perspective of experiencing and reflecting performance. However, students have to be able to experience that various sports can be exerted with an emphasis on diverse perspectives, even those that seem to be bizarre at first sight (cf. Bräutigam, 2011a, p. 86). These experiences students gain from the multi-perspective view of physical education are crucial since they – partly – account for the legitimation of school sport. Moreover, those experiences are subject to students' attitudes, as well as to their perception.[20] Therefore, it seems to be worthwhile to investigate students' attitudes toward the latter to get an insight into how students evaluate and perceive the crucial experiences they should gain from physical education classes.

As shown in this section, the elements of the physical education class as well as the conception of the latter are the main constituents of what we understand as a physical education class. Moreover, they are direct subject to students' evaluation and perception

---

[19] cgs = centimeters, grams, seconds.

[20] In chapter 3.2, the implementation of these experiences in the present study is discussed in more depth.

and thus crucial for this study. In section 3, the choice of constituents will be further constrained and explained.[21]

2.1.4 Physical Activity and Exercise

Beside the key terms that are directly intertwined with the present investigation, another term is important to define. The questionnaire used in this study contains a passage concerned with the students', the parents' and the students' best friends' amount of exercise and physical activity. This is important since one of the underlying hypothesis is that the amount of exercise and physical activity influences the perceptions and attitudes toward physical education classes. A distinction between physical activity and exercise is therefore important to sort out this section of the questionnaire and the objects that are connected to it.

Physical activity has been defined as "any bodily movement produced by skeletal muscles that requires energy expenditure" (World Health Organisation, 2010, p. 53). Jackson, Morrow, Hill and Dishman (2004, p. 4) share this definition, determining that "in simplest terms, physical activity means moving about". These vague statements indicate that almost every movement of the body is considered to be physical activity. These definitions are further specified as physical activity constituting "any bodily movement produced by skeletal muscles that result in energy expenditure beyond resting expenditure" (Thompson, Buchner, Pina, Balady, Williams, Marcus, Berra, Blair, Costa, Franklin, Fletcher, Gordon, Pate, Rodriguez, Yancey, & Wenger, 2003, p. 1). Consequently, physical activity is not a state of complete rest, but rather an active motion. Moreover, physical activity positively correlates with physical fitness (cf. Caspersen, Powell, & Christenson, 1985, p. 127). This further restricts the range of the

---

[21] The purpose of this study is not, however, to profoundly investigate the subject of school sport. Therefore, a detailed discussion of the interrelations of all the fields of didactic and pedagogy is omitted here, since this would go beyond the scope of the discussion.

discussed term. Physical activity is, however, not to be equated with exercise. In fact, exercise is rather a "subset of physical activity that is planned, structured, repetitive, and purposeful [...]" (Thompson et al., 2003, p. 1).

The difference between the two terms at hand therefore is the exercise's planned and structured nature. As Thompson et al. (2003) and Caspersen et al. (1985, p. 127) also incorporate this structured and planned element in their definition. Furthermore, they also stress the objective to improve one's physical fitness and assign exercise a very positive correlation with physical fitness – whereas physical activity is only assigned a positive correlation with physical fitness.[22]

It is therefore important to distinguish between physical activity and exercise – considering the latter to be a subcategory of the prior – and to be aware that even simple activities such as doing dishes or brushing ones teeth can be seen as physical activities. Figure 4 depicts an attempt of differentiation of the two concepts.

By the same token, this study follows Jackson et al.'s (2004, p. 4) approach that "there are, of course, degrees of physical activity". This makes it possible to restrict the term of physical activity to actions that involve more than just one skeletal muscle and in terms of duration. On top of this, following Caspersen et al. (1985), the positive correlation with physical fitness and the increased energy expenditure allows the study to limit physical activity to active bodily work, such as gardening, walking to school or riding one's bike. The implementation of the just discussed terms into the questionnaire will be further explained in chapter 3.

---

[22] Interestingly enough, however, as Vanhees, Lefevre, Philippaerts, Martens, Huygens, Troosters, and Beunen (2005) indicate, physical fitness is not only dependent on physical activity or exercise, but also on other attributes. See Vanhees et al. (2005) for further information.

Elements of physical activity and exercise	
PHYSICAL ACTIVITY	**EXERCISE**
1. Bodily movement via skeletal muscles	1. Bodily movement via skeletal muscles
2. Results in energy expenditure	2. Results in energy expenditure
3. Energy expenditure (kilocalories) varies continuously from low to high	3. Energy expenditure (kilocalories) varies continuously from low to high
4. Positively correlated with physical fitness	4. Very positively correlated with physical fitness
	5. Planned, structured, and repetitive bodily movement
	6. An objective is to improve or maintain physical fitness component(s)

Figure 4: Differentiation between Physical Activity and Exercise (Source: Caspersen et al., 1985, p. 127).

2.2 Importance of Physical Activity and Physical Education

In the preceding chapter, key terms that are crucial for the understanding of the present study have been defined. This review of literature concerning those key terms, however, still does not include the qualitative and quantitative research that has been conducted in the fields of physical activity and physical education.

Many studies in the mentioned domain have been realized, distinguishing themselves in the focus on different topics. Primarily, physical education students' self-perceptions have been investigated (cf. e.g., Kirkcaldy, Shephard, & Siefen, 2002). Besides, the health behaviors of physical education students have been a field of interest (cf. e.g., Pate, Heath, Dowda, & Trost, 1996).

The following sections will therefore provide a review of current studies, lining out the importance of physical activity, exercise and physical education as already mentioned. Correspondingly, the most important findings of research related to physical activity and physical education classes will be presented. The purpose is not, however, to present a complete demonstration of the state of research in detail, but rather to present selected findings that are related to the topic of this study.

2.2.1   Health Behavior and Physical Education

The dual function of physical education classes comprises an education by the means of sport on the one hand, and an education toward sport on the other hand. The question of why children are to be educated to engage in sport on a regular and – in the best case – lifelong basis, is often answered in terms of health.

Moreover, Hardman (1998) stresses the role of physical education in promoting healthy well-being to enrich the quality of life. Since this promotion of health constitutes a central factor of physical education classes' content, and the content and the outcomes of physical education classes are subject to attitude formation and closely related to the present study, it is important to line out current investigation examining the relation between exercise and physical education on the one hand and health on the other hand.

The effects of exercise and physical activity on health are well known. Harris and Cale (1997) point out exercise recommendations for adolescents, including enjoyable sports and games comprising a large group of skeletal muscles as well as an active lifestyle, such as cycling to school.

In the context of the *2. Deutscher Kinder- und Jugendsportbericht*, Schmidt, Zimmer, and Völker (2009) report that two thirds of 10 to 13 years old children's afternoon activities fall upon sports. However, these physical activity levels decline when children trespass to the crucial period of adolescence. These observations are further supported by the findings of the US Department of Health and Human Services (2000) that reports that most US people including adolescents and children do not meet the recommendations for physical activity.

Various authors have set these recommendations for adolescents and children to at least an hour of physical activity almost every day (cf. Cavill, Biddle, & Sallis, 2001; Prochaska, Sallis, & Long 2001). Despite these facts, health has been reported to be a major reason for both adults and adolescents to engage in physical activity (cf. Pano & Markola, 2011, p. 60),

Long-since, the health related benefits of physical activity and exercise have been included in the goals of physical education classes through health-related exercise (cf. Green & Lamb, 2000, p. 88). Furthermore, as already indicated in chapter 2.1, health and the development of a health consciousness constitute one of the six sense perspectives of multi-perspectival physical education classes.

Interviewing teachers in North-West England, Mason (1995, p. 6) found out that most teachers "stressed the health-related benefits of sport and exercise for children". Students are to be encouraged by the means of physical education to engage in an active lifestyle throughout life (cf. Green & Lamb, 2000, p. 89). Thus, sedentary behavior can be reduced and adolescents – in the present and in the future – are able to benefit from the health related advantages of physical activity and exercise (cf. Reilly, Penpraze, Hislop, Davies, Grant, & Paton, 2008, p. 614).

Biddle Sallis, and Cavill (1998) further highlight the significance and potential of physical education classes to help children reach their recommended amount of physical activity. The view of educating students toward physical activity and sport – of course – concurs with the dual function of physical education classes established in German scholarly literature as described above. Simons-Morton (1994) emphasizes this point, as well, assigning school sport the task of giving students the possibility to engage in sufficient amounts of physical activity on the one hand, and equip them with knowledge and skills to be physically active outside school on the other hand.

It is important to mention, however, that this enhanced emphasis on health in physical education classes has not led to a rejection of traditional sports and other physical education class contents (cf. Roberts, 1995, p. 339).[23] Moreover, a longitudinal study on adolescents and the development of their physical activity levels revealed that adolescents reduce their levels of physical activity and exercise in the crucial phase from 12 to 15 (cf. Aaron, Storti, Robertson, Kriska, & LaPorte, 2002, p. 1075). This agrees with the findings of the *2. Deutscher Kinder- und Jugendsportbericht,* as mentioned above.

The interesting – but yet unsolved – question is, if this decline concurs with negative attitudes toward physical activity and exercise – and maybe toward physical education. Other investigation in the field of health, school sport, and physical education have shown that participation and activity in physical education classes are dependent on several individual factors. Students with higher body fat tend to be less active, highly skilled students are more active than their (lowly skilled) peers and boys tend to be more active than girls (cf. Fairclough & Stratton, 2005, p. 15).

Since one of the functions of physical education classes is to educate students toward physical activity and exercise, the above findings are crucial to the conceptualization of school sports. School sport has to find a way to delight adolescents and to enthuse them

---

[23] These findings suggest that many teachers use the health benefits of physical activity and exercise as legitimation for physical education classes. The consequential question therefore is, whether the subject of health is stressed implicitly or explicitly, that is, whether this subject is reflected and discussed with students to make them aware of health benefits. The maintenance of the classic physical education class contents is not necessarily indicative of implicit incorporation of the subject of health. The sense perspectives are rather to be experienced by the means of various contents. Yet, a closer examination of this subject is not the aim of the present study.

with physical activity and sport. One way to influence students in the intended way may be in understanding their attitudes toward the subject and thus the physical education teachers' influence on those attitudes.

2.2.2 Students' Self-Perception in Physical Education

Besides the benefits related to physical health, physical activity and exercise have also been reported to contribute to psychological well-being, including stress-management skills, positive mood and an enhanced self-concept (cf. Kirkcaldy & Shephard, 1990). The self-concept has been a broadly investigated field in psychology, and its value for schools and educational contexts has frequently been emphasized (Marsh & Craven, 2006). It is defined as a "cognitive and structured unit, a perception of who we are[24] – e.g., the conviction of being a good basketball player" (Woolfolk & Schönpflug, 2008, p. 108).[25] This definition suggests the assumption that self-concept and self-perception are closely related to participation in, and motivation and attitude toward physical activity, exercise and physical education (cf. Page, Ashford, Fox, & Biddle, 1993, p. 585). This idea coincides with Fairclough and Stratton's (2005, p. 15) view of skilled students being more active, as already pointed out in chapter 2.2.1. Fox (1998) emphasizes the impact of the physical self-perception on the global self-concept, which stresses its role in physical education classes and vice versa. Figure 6 displays a further segmentation of the physical self-perception into the four subdomains *Sport Competence*, *Body Attractiveness*, *Physical Strength*, and *Physical Condition*.

---

[24] Cf. Weiten, Dunn, and Hammer (2012).

[25] Translated from German, the original passage quotes as follows: "Selbstkonzept ist eine kognitive, strukturierte Einheit, eine Vorstellung davon, wer man ist – z.B. Die Überzeugung, dass man ein guter Basketballspieler ist." Moreover, Woolfolk, and Schönpflug (2008) further divide the global self-concept in subcategories, such as the ability self-concept or the social self-concept. However, for the purpose of this study, the above noted definition is sufficient. For further information see Woolfolk and Schönpflug (2008).

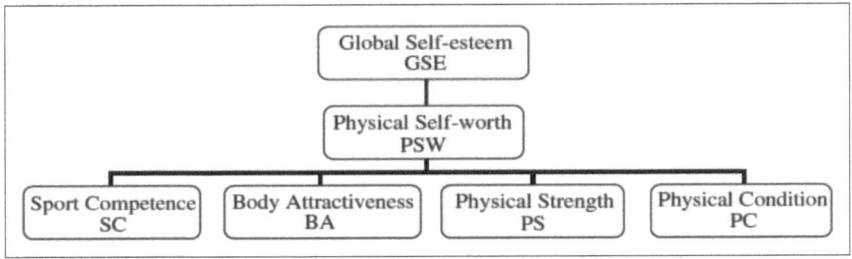

Figure 5: The Physical Self-Perception and its subdomains (Source: Fox & Corbin, 1989, as printed in Raustorp et al., 2005, p. 127).

Furthermore, Woolfolk and Schönpflug (2008, p. 108) state that from childhood to adolescence, the self-concept is closely related to physical appearance, performance and social acceptance, and that it is subject to social comparison. This again supports the idea of self-concept and self-perception influencing the field, which is of interest of this study.

Many studies examining the physical self-perception have been conducted. As a main instrument, the PSPP – physical self-perception profile – has been used to investigate this field in the context of physical activity and physical education. Marsh (1998) notes that physical self-perception significantly declines in the adolescent years from 12 to 16. Besides, males are reported to have better scores in physical self-perception than females (cf. Maiano, Ninot, & Bilard, 2004, p. 64).

These two findings concerning age and gender may be due to the contents of physical education classes or due to Western culture's view and perception of the body. However, there has also been evidence for a positive correlation between participation in sport and self-perception (cf. Kirkcaldy et al., 2002, p. 548). This study suggests that those adolescents who regularly participate in physical activity and exercise have a better image of themselves. In addition to that, Kirkcaldy et al. (2002, p. 548) also point out the associations between regular physical activity and exercise and attitudes. Other studies have focused on the correlation between physical self-worth and sport competence (cf. Raustorp, Stahle, Gudasic, Kinnunen, & Mattsson, 2005, p. 132). Higher skilled students are reported to have a higher physical self-worth and therefore a better self-perception than lower skilled students. This again impacts the idea of skilled students being more active in physical education classes and thus having better attitudes toward the subject (cf. Fairclough & Stratton, 2005). An implication for school sport therefore is to teach children and adolescents a variety of skills to improve their physical self-perception and thus their behavior in and attitude toward physical education classes.

Even if a relation between physical self-perception and attitudes toward physical activity and exercise, and physical education has been indicated in this chapter, this field needs to be examined further to reveal the impacts those two concepts have on each other in physical education classes.

2.3 Current Research on Physical Education Classes

The importance of physical activity, exercise and physical education has been lined out in the preceding chapter. Even though the benefits of physical activity, exercise and physical education are almost indisputable, the issue of students' attitudes toward physical activity, exercise and physical education has not been solved in this way. In the existing literature, the motivations of students have been examined (cf. e.g., Biddle & Wang, 2002), and researchers have – as in the present study – focused on the perceptions and attitudes of students towards physical education (cf. e.g., Stelzer, Ernest, Fenster, & Langford, 2004). Moreover, some research related to German physical education classes exists, as well (cf. e.g., Brettschneider & Kuhlmann, 2006).

This chapter will give a summary of selected existing studies and corresponding findings that are related to the subject of the present study, that is attitudes and perceptions of students toward physical education.

2.3.1 Motivation and Physical Education

Even though the benefits of physical activity, exercise and physical education are almost indisputable (Bailey, 2006; U.S. Department of Health and Human Services, 2008), the issue of students' attitudes towards physical activity, exercise and physical education has not been solved in this way. In the existing literature, the motivations of students have been sufficiently examined (Biddle & Wang, 2003), and researchers have also focused on the perceptions and attitudes of students towards physical education (Stelzer, Ernest, Fenster, & Langford, 2004).

As attitude, motivation has been a highly examined field of social psychology, and the relation between the two concepts at hand is not distinctively exposed, but "attitudinal factors are shown to directly underlie motivational attributes [...]" (Gardner, 1985, p. 10). On the other hand, the difference between motivation and attitude is said to be their orientation: whereas attitudes are – as explained above – object orientated, motivation is directed against a specific goal (cf. Utsch, 2007, p. 56). Yet, the important finding is that – despite the blurred relation of both concepts – motivation has a major influence on behavior (cf. Gage & Berliner, 1996, p. 339; Locke & Braver, 2008) – just as attitudes do.

In investigating the role motivation plays in physical education classes as well as in other educational fields, many researchers have adopted Deci and Ryan's (1985) *Self-Determination Theory* to motivation. This approach highlights the importance of different types of motivation[26], and especially the importance of the self-determined behavioral regulation in the field of physical education has been shown (cf. Wang & Biddle, 2001). The importance of self-determined motivation in educational contexts

---

[26] Although it is not subject of this study, it is interesting to remark the different types of motivation. Deci and Ryan (1985) differentiate between intrinsic and extrinsic motivation, where extrinsic motivation is further subdivided into external regulation, introjected regulation and identified regulation. Those different types of motivation form a continuum. For further information see Deci and Ryan (1985).

was further shown by Vallerand, Pelletier, Blais, Briere, Senecal, and Vallieres (1993), revealing a relation to grading, classroom atmosphere and future learning intentions. In the context of sports, similar findings are reported by Pelletier, Fortier, Vallerand, Tuson, Briere, and Blais (1995). Intrinsic motivation was positively related to effort and future intentions (cf. Pelletier et al., 1995, p. 48).

A further modification of the self-determination approach with interesting implications for physical education classes and this study presents Vallerand (1997), stating that various social factors influence the different types of motivation. Physical activity and physical education exhibit several social factors that in turn influence and determine student motivation toward the physical education class. Among the most important factors the interaction between students, the physical education teacher's emphasis on self-referenced improvement or the possibility for students to choose tasks (cf. Ntoumanis, 2001, p. 227).[27] This is especially interesting in the context of the present study, since interaction is one of the main components of the questionnaire used for this investigation and therefore forms a junction between the concepts of attitude and motivation.[28] These social factors are further defined in terms of a person's psychological needs: the satisfaction of the needs for autonomy, competence and relatedness will therefore facilitate self-determined behavioral regulation (cf. Ntoumanis, 2001, p. 228). More recent research has confirmed this concept for the fields of exercise and physical education, however placing more emphasis and importance on competence and relatedness (cf. Standage, Duda, & Ntoumanis, 2003, p. 106). This further underlines the relation between motivation and attitude. Figure 7 displays the self-determination continuum.

---

[27] Vallerand and Losier (1999, pp. 145ff)) use the terms "competition and cooperation", "success and failure" and "the coach's behavior toward the athlete" for the above described social factors. However, they can easily be transferred to the context of physical education classes.

[28] See chapter 3 for further information about the choice of items concerning the questionnaire.

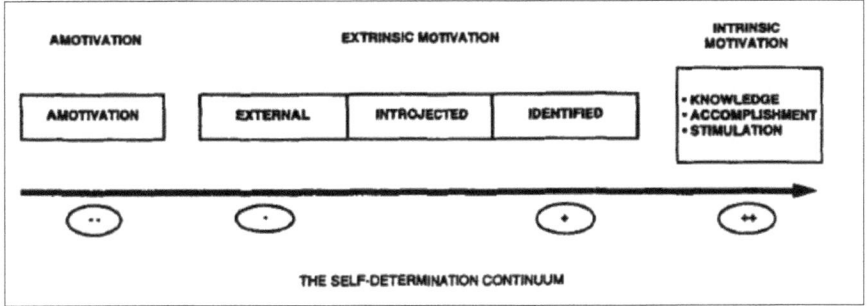

Figure 6: The different Types of Motivation according to the Self-Determination Theory (Source: Deci & Ryan, 1985, as printed in Vallerand & Losier, 1999, p. 153).

2.3.2 Research on German Physical Education Classes

Physical education is a field that has been subject to a lot of research and scholarship also in Germany, and recent German studies more and more focus on the actual processes of physical education classes. Many reports and investigations are concerned with the general development of school sports (cf. e.g., Serwe & Thiele, 2008), the curricula of physical education classes (cf. e.g., Stibbe, 2011), or with the issue of teaching physical education classes and the associated problems (cf. e.g., Kastrup, 2009).

However, besides the teachers, the central persons affected and the main constituents of physical education classes are the students, playing a decisive role in constructing physical education classes (cf. Bräutigam, 2011b, p. 65). Thus the design of physical education classes is to be oriented toward the students, and it is no coincidence that the first aspect that is to be taken into account in the process of planning lessons are the students (cf. Döhring & Gissel, 2009).

Nevertheless, only few recent German publications are concerned with the issue of students and their perceptions and attitudes of, and needs and wishes in the physical education classes. In this subchapter, some selected findings of German investigation in the field of physical education and physical activity with a main focus on the subject of students that seem to be important to the examination of students' attitudes and perceptions are reflected.

Some studies have completely focused on the sportive living environment of children and adolescents. Even though these findings do not directly refer to physical education classes, they are crucial to investigate the physical behavior of children and adolescents, which in turn might be indicative of their attitudes. According to the KiGGS[29], the majority of adolescents at the age of 11 to 17 participate in physical activity or exercise at least once a week. One third of the males, and more than half of the females, however, do not engage in such activities more than two times a week, and thus do not meet the in chapter 2.2.1 outlined recommendations of physical activity and exercise (cf. Lampert, Mensink, Romahn, & Woll, 2007, p. 640).

The report further indicates that physical activity and exercise decline in the course of adolescence, and that multiple gender-specific differences exist, such as the influence of migration or place of residence. Whereas these factors do influence females, males do not seem to be affected (cf. Lampert et al., 2007, p. 639). The study further points out recommendations to promote physical activity and exercise, yet it does not comprise physical education and school sport as possible point of application. Figure 8 represents the described findings of adolescents' sporting activity.

---

[29] KiGGS is the abbreviation of the *Kinder- und Jugendgesundheitssurvey* (German Health Interview and Examination Survey for Children and Adolescents), a nationwide research executed between 2003 and 2006 by the Robert-Koch-Institut Berlin to investigate physical activity and exercise among children and adolescents.

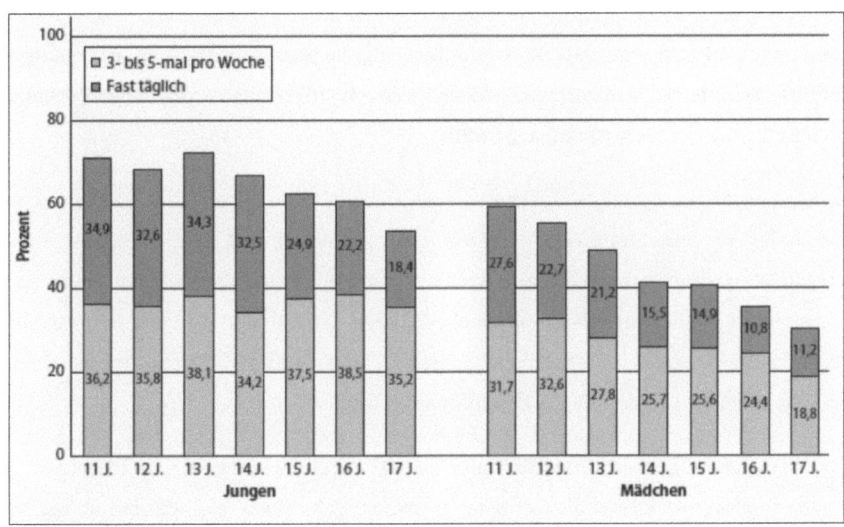

Figure 7: Sporting activity of adolescents', separated according to gender (Source: Lampert et al., 2007, p. 639).

Another study investigating the potential of physical activity, exercise and physical education in an extracurricular environment displays the *2. Deutscher Kinder- und Jugendsportbericht*, with the purpose to illustrate the benefits of physical activity and exercise on the development of children and adolescents (cf. Deutsche Sportjugend, 2009, p. 1).

As already indicated in chapter 2.2.1, almost two thirds of children and adolescents' afternoon activities fall upon physical activities and exercise. Physical activity and exercise is said to positively affect behavior and state of health both in the present and in the future, and physical activity and exercise are reported to contribute to children's and adolescents' sense of self-worth (cf. Gieß-Stüber, Neuber, Gramespacher, & Salomon, 2009, p. 67). These findings are in accord with the outcomes of international studies outlined in the preceding chapters and further validate the importance of physical activity and exercise in the education of children and adolescents. Furthermore, the investigation presents data directly influencing the process of learning. Accordingly, an early facilitation of motoric skills does positively affect the motoric development as well as the development of language skills.

In addition, the study displays a positive influence of physical activity and exercise on the ability to concentrate (cf. Rethorst, Fleig, & Willimczik, 2009, p. 253). These outcomes constitute more reasons to justify physical education classes and to investigate the attitudes of students toward physical education to be able to design a best possible learning environment for this crucial subject. Correspondingly, Schmidt et al. (2009) present guidance to stress physical activity even further.

Beside the above outlined extracurricular studies, research concerned with the subject of physical education classes in Germany exists, as well. Since, as indicated above, the students are the main subjects of physical education classes, teachers have to be aware of their perspectives, beliefs and sentiments. Bräutigam (2011b) summarizes central findings concerning the perception of students toward physical education. Accordingly, school sport is both important and interesting for a majority of students, and performs very well in popularity ratings (cf. Bräutigam, 2011b, p. 72). Moreover, students are reported to be very contented with physical education classes, to have fun during classes and to feel well after classes. Interestingly enough, this satisfaction does also hold true for students' perceptions of teachers and their behavior (cf. Bräutigam, 2011b, p. 73).

Beside these positive findings, Bräutigam (2011b) also portrays the reverse side of the state of research. As a matter of fact – beside the majority that positively evaluates physical education classes – a considerable amount of students exist who negatively assess school sport. Furthermore, the popularity of physical education classes distinctly declines in the course of school years (cf. Bräutigam, 2011b, p. 74).

These outcomes of field research display interesting facts, which do also play a role for the present study.[30] Bräutigam's (2011b) explanation also incorporates most student-related findings of the *DSB-SPRINT Studie* (DSB, 2006). The purpose of this study conducted by several German study groups is to investigate school sport with regard to all conditions that influence its realization. As a result, the main examination objects are the situation of sport facilities, extracurricular sport and the perspectives of teachers, students, parents and school administration (cf. Brettschneider & Kuhlmann, 2006, pp. 4ff).

Beside the already mentioned student-related outcomes, the study emphasizes the students' wishes to execute more current trend sports and the insufficient adjustment of school sports with regard to students needs (cf. Gerlach, Kussin, Brandl-Bredenbeck, & Brettschneider 2006, pp. 114ff). It seems, however, that the *DSB-SPRINT Studie* (DSB, 2006) boldly emphasizes negative aspects of school sports, neglecting all the positive findings.[31] For this reason, the outcomes of the *DSB-SPRINT Studie* (DSB, 2006) have to be further examined and validated to provide useful guidance.

In another qualitative study of perceptions in physical education classes, Krieger (2006) opposes teacher perceptions and those of the students. The investigation suggests that it is indispensable to realize the gap between the teacher's perception and knowledge about a situation on the one hand, and the students' perception and knowledge on the other hand (cf. Krieger, 2006, p. 69), and that further research in this field is required.

---

[30] Some modules of these findings are also incorporated in the questionnaire at hand. See chapter 3 for further information and clarification.

[31] The Ministerium für Kultus, Jugend und Sport (Ministry for Cultural Affairs, Youth, and Sport) Baden-Württemberg (2005) expounds this and other critical issues concerning the execution of the *DSB-SPRINT Studie* and the application of the outcomes on single states. See Ministerium für Kultus, Jugend und Sport (2005) for further information.

Its focus on students' perceptions makes the KIKSS[32] an interesting survey in the context of the present study. The results of this investigation suggest a huge impact of teachers' behavior on students' enjoyment of physical education classes (cf. Heemsoth & Miethling, 2012, p. 10). This is an interesting finding, since – as will be shown in chapter 2.3.3 – teachers' behavior is said to impact students' attitudes, as well. Besides, from a general point of view, it seems obvious to suggest that high enjoyment of physical education classes positively affects attitudes. However, further validation of the presented construct is indispensable to solve these relations, since the questionnaire has not been entirely validated.

Also in the context of student-orientation, Volkamer (1997, p. 52), on the basis of his observations in day-to-day life and physical education classes, points out a central dilemma of physical education, stating:

> "If we on the one hand know that in the rarest cases students find their ways to club sports through physical education classes, but that, on the other hand, physical education classes often spoil sport for those who really need it, we can say: for good athletes, physical education classes are at best a pleasant addition, but actually unimportant. For bad athletes, however, it possibly is an additional cause for refusal of sport, and hence important."[33]

---

[32] KIKS is the abbreviation for *Kieler Fragebogen zum Unterrichtsklima im Sportunterricht aus Schülersicht* (Questionnaire of students' perceptions toward the class climate in physical education classes). This questionnaire is a first approach to investigate the factors affecting the class climate in physical education classes. See Heemsoth and Miethling (2012) for further information.

[33] Translated from German, the original passage quotes as follows: "Wenn man nun einerseits weiß, dass Schüler in den seltensten Fällen über den Schulsport in den Verein kommen, dass aber andererseits die Schule denjenigen häufig den Sport vermiest, die ihn am nötigsten brauchen könnten, kann man sagen: Für die guten Sportler ist der Sportunterricht allenfalls eine erfreuliche Zugabe, aber eigentlich unwichtig, für die schlechten Schüler dagegen ist er möglicherweise ein (zusätzlicher) Ablehnungsgrund – also wichtig." (Volkamer, 1997, p. 52).

The statement concurs with the findings of other recent research (cf. Hartmann-Tews & Luetkens, 2003). This dilemma reveals the necessity of a profound understanding of students' perceptions and points of view to succeed in teaching physical education classes. Without an understanding of how the students think – especially the "bad athletes" – it will not be possible to overcome this central problem of physical education classes.

2.3.3 Students' Attitudes toward Physical Education

It appears hard to strictly distinguish between research that is solely concerned with attitudes from approaches dealing with perceptions, experiences, beliefs, opinions, and/or views. On the item level in quantitative instruments or on the question level in qualitative research, phrasing and perspectives may be similar or even identical in some cases (Carlson, 1995; Ennis, 1996; Portman, 1995). However, the following comprehensive display of research findings in the field of students' attitudes towards physical education will try to cover only relevant studies that have explicitly focused on the construct "attitude" or have argued based on a scientific framework that integrates attitude as a prominent construct. The focus groups of college students and teachers will also not be covered, since age group and context are essentially different compared to school. However, one should note that there is a growing body of research concerned with college students' and physical education teacher' attitudes (Goktas, 2012; Keating, Silverman, & Kulinna, 2003; Sanes, 2009).

Especially on an international level, attitudes towards physical education have been subject to research and field studies in the last two decades. Attitude has frequently been researched as dependent variable (Silverman & Subramaniam, 1999), and thus, focus in this paragraph is laid mostly on the independent variables influencing attitude.

With the aim to examine students' attitudes and perceptions towards physical activity and exercise, Pano and Markola (2011) developed a questionnaire with several questions related to participation in extra-curricular physical activity and exercise. This questionnaire was developed to determine the degree of physical activity and exercise of adolescents. Besides, reasons for engagement and non-engagement in physical activity and exercise were inquired. Main findings of this study were that students participated both in physical activity and exercise with some regularity - that is at least three times a week -, that fun, health improvement and improvement of physical performance are the major reasons for participation, and that a lack of time detained them from more participation in these activities.

In terms of gender, many studies suggest that boys are physically more active than girls (Armstrong & Bray, 1991; Crespo et al., 2013; Sleap & Warburton, 1992), and that men are more active than women (Suminski, Petosa, Utter, & Zhang, 2002). Unexpectedly, however, the results of a field study by Birtwistle and Brodie (1991) attributed girls more positive attitudes towards physical activity than boys. By the same token, however, girls were assigned more negative attitudes towards physical education classes, which may be due to the contents of physical education classes. Yet, these outcomes are disagreed in more recent studies, stating that boys show more positive attitudes towards physical activity and the scientific benefits of exercise (Arabaci, 2009; Omar-Fauzee et al., 2009).

The different outcomes may be due to different cultural environments - as suggested in studies involving students from different countries (Chung & Phillips, 2002; Stelzer et al., 2004; Tannehill & Zakrajsek, 1993) - or due to different grade levels. Adding to the mixed results, Ilker, Arslan, and Demirhan (2011) did not find differences in students' attitudes in regard to gender and grade level, whereas AL-Liheibi (2008) did find more positive attitudes towards physical education in high school students than in middle school students. Other researchers stress the structure of physical activity and exercise. Whereas boys tend to favor challenge and risk, girls tend to prefer aesthetics (Silverman & Subramaniam, 1999; Subramaniam & Silverman, 2002). These differences may also account for the gender related differences in attitudes. Nevertheless, these contradictory findings obviously require more research regarding the attitudes of students towards physical activity and exercise in relation to various independent variables.

In the context of physical education, the National Association for Sport and Physical Education (2004) states that interest in physical activity and physical education classes declines in the course of adolescence. These outcomes have been supported by various other field studies as well (Prochaska, Sallis, Slymen, & McKenzie, 2003; Subramaniam & Silverman, 2007). A field study on British students' perceptions revealed that students considered physical education to be an important subject, and that alongside with it, these perceptions were not dependent on socio-economic status (Birtwistle & Brodie, 1991). These findings are further supported by Valdez (1997). In Valdez's (1997) study, no relation between the dependent variable attitude towards physical education and the independent variables gender, ethnicity and socio-economic status are found. On the contrary, Tannehill and Zakrajsek (1993) found differences according to ethnic group, but the sample was not representative and no inferential statistical procedures were applied.

Supporting Birtwistle and Brodie's (1991) results, a field study by Anderssen (1993) revealed that boys have more positive attitudes towards physical education than girls. In line with these results, Chung and Phillips (2002) also reported generally higher scores for attitudes towards physical education in boys than in girls. Anderssen's (1993) study further illustrates that those students who are physically more active, perceive physical education classes more positively. In turn, Chung and Phillips (2002) reported a significant relationship between students' attitudes and their leisure-time exercise. This mutual relationship may be due to the fact that skilled students gained more positive experiences than their lower skilled peers (Silverman, 1996). This interesting finding is reinforced by several other studies that are concerned with the influence of skill level on perception and attitude (Ennis, 1996; Portman, 1995). Haynes et al. (2008) found that students very well appreciated grouping according to individual skill levels in physical education. Another striking finding is that students' attitudes and perceptions towards physical education are furthermore influenced by the teacher and the curriculum. In a qualitative study, Luke and Sinclair (1991) identify the teacher as a major determinant of students' attitudes and perceptions.

Solmon and Carter (1995) report that a variety of activities in physical education classes result in more positive attitudes towards the subject, and Rikard and Banville (2006) emphasize the importance of content variability to improve students' attitudes. This further confirms the impact of physical education content on students' attitudes (Bibik et al., 2007). The possibility of implementing attitude-changing strategies - partly through variation of physical education class content - as suggested by Theodorakis and Goudas (1997) is shown in a longitudinal field study. In a one-year intervention in seventh grade physical education classes, an attitude change is reported after applying several strategies to alter attitudes towards physical activity and exercise, and physical education (Digelidis, Papaioannou, Laparidis, & Christodoulidis, 2003). The understanding of attitude formation, and knowledge about actual attitudes of students may help physical education teachers to perform physical education classes in a way to facilitate more positive attitudes towards physical education as well as towards physical activity.

In sum, accumulated evidence suggests that individual characteristics (such as age, gender, and sports skill) and contextual factors (such as physical education curriculum, comprehensive intervention or physical education programs, organized sports programs, and physical education teachers) impact students' attitudes towards physical education and physical activity (Xu & Liu, 2013).

Including students and their parents, Tannehill, Romar, O'Sullivan, England, and Rosenberg (1994) examined students' attitudes as well as parental attitudes towards physical education. Results indicated mixed attitudes towards physical education in both target groups. However, parents did not attribute physical education's value with lifelong physical activity, which is alarming from a physical education pedagogy perspective.

Applying a qualitative approach that targets certain components within physical education, Bernstein, Phillips, and Silverman (2011) interviewed middle school students about their attitudes and perceptions towards competitive activities in physical education. Their study's findings suggest that students' attitudes are impacted by the structure of the competitive activities themselves. Certain experiences such as structuring teams, evolving team leaders, selfish game-play, and excessive losing or winning behavior effect students' attitudes towards competitive activities.

In a sophisticated research design, Dismore and Bailey (2011) investigated the effects of fun and enjoyment in physical education on student's attitudes towards it. Conducting a two-phase study, they found that fun and enjoyment featured prominently in students' reports on attitudes towards physical education. However, during transition from Key Stage 2 (7- to 11-years old) to Key Stage 3 (11- to 14-years old), fun appears to be reinterpreted by the students. Students in Key Stage 3 associated fun more with learning, thereby deeming physical education and fun in it a valuable learning experience.

Overall, students' attitudes towards physical education appear to be mixed. Although the majority of the studies report positive attitudes towards physical education (AL-Liheibi, 2008; Chatterjee, 2013; Colquitt, Walker, Langdon, McCollum, & Pomazal, 2012; Mohammed & Mohammad, 2012; Rikard & Banville, 2006; Ryan et al., 2003; Subramaniam & Silverman, 2002; Zeng et al., 2011), some studies report mixed results in detail (Bernstein et al., 2011; Ilker et al., 2011; Tannehill et al., 1994), and some studies reveal negative attitudes towards physical education (Carlson, 1995; Orunaboka, 2011; Portman, 1995). Nonetheless, physical education still seems to be one - maybe the most - liked and appreciated subject from the student's perspective (Bibik et al., 2007; Coulter & Woods, 2011).

Table 1 displays an overview of the relevant studies on student's attitudes towards physical education, including brief research outcomes and focus groups.[34]

Table 1: Overview of Studies on Students' Attitudes toward Physical Education.

Study	Focus Group	Methodology	Major Outcomes
AL-Liheibi (2008)	Middle and high school students (N=480)	Quantitative (survey)	1) High school students had more positive attitudes than middle school students (personal satisfaction). 2) Students with gym access had more positive attitudes than students without gym access. 3) Students engaging in daily physical activity outside school had more positive attitudes towards physical education than students who engaged in less physical activity.
Arabaci (2009)	Secondary school students (N=1240)	Quantitative (questionnaire)	1) Male students' attitudes changed reached higher scores than female students. 2) Younger students preferred co-educated classes, whereas older students preferred single-sex physical education classes.
Bernstein et al. (2011)	Middle school students (N=24)	Qualitative (interviews)	Three major themes emerged: 1) Having fun in competitive activities 2) Not all students were attaining motor skills necessary to participate (due to lack of time to engage in appropriate practice). 3) The structure of competitive activities affects student experience.
Bibik et al. (2007)	High school students (N=223)	Quantitative (survey)	1) Students would enjoy having more sports or games in their physical education curriculum. 2) Students indicated that physical education was important to their high school education. 3) Students who indicated they enjoyed physical education were more likely to enjoy school. 4) Students who engaged in negative health behaviors were less likely to enjoy physical education.

---

[34] Cf. Kretschmann and Wrobel (2014), and Kretschmann (2015).

Birtwistle & Brodie (1991)	Secondary (N=291) and primary school (N=316) students	Quantitative (questionnaire)	1) Girls had significantly more positive attitudes than boys. 2) The data yielded no differences in attitudes between socio-economic levels.
Chatterjee (2013)	Secondary school students (N=273)	Quantitative (questionnaire)	1) Participants had positive attitudes towards physical education. 2) No significant differences in gender and geographical location were found.
Chatterjee et al. (2012)	Secondary school students (N=273)	Quantitative (questionnaire)	1) Students that are highly motivated in terms of sports perfectionism are more likely to show positive attitudes towards physical education. 2) Parental pressure may increase attitude decreases in their children.
Chung & Phillips (2002)	High school students (N=451)	Quantitative (questionnaire)	1) Significant relationships between attitudes towards physical education and leisure-time exercise were found, regardless of gender or nationality. 2) Significant differences were found in attitudes towards physical education by gender and nationality.
Colquitt et al. (2012)	Middle school students (N=122)	Quantitative (survey)	1) Overall, students had positive attitudes towards physical education. 2) PACER test was the only significant predictor of enjoyment in physical education. 3) PACER and BMI were significant predictors of perceived usefulness of physical education.
Dismore & Bailey (2011)	1) Key Stage 2 (7- to 11 years old) students (questionnaires: N=790; interviews/ focus group: N=86) 2) Key Stage 3 students (7- to 11 years old) (questionnaires: N=875; interviews/ focus group: N=10)	Mixed method (questionnaire, focus group, interviews)	1) Fun and enjoyment featured prominently in reports on attitudes towards physical education. 2) For Key Stage 2, fun appeared to be a critical factor for making physical education enjoyable. 3) Following transition (Key Stage 2 to Key Stage 3), many children began to describe fun in terms of learning, indicating that they valued physical education more than before.
Haynes et al. (2008)	High school students (N=67)	Qualitative (interviews)	A number of participants described a change in attitude towards physical education attributed to a grouping arrangement based on individual skill.

Ilker et al. (2011)	High school students (N=1604)	Quantitative (survey)	1) Turkish high school students' attitudes towards physical education were neutral. 2) Attitudes did not change according to gender and grade level.
Ilker & Demirhan (2013)	High school students (N=81)	Quantitative (questionnaire)	Students exposed to a mastery-oriented motivational climate show significantly improved scores for attitudes towards physical education, compared to students who were exposed to a performance-oriented or a performance avoidance-oriented motivational climate.
Luke & Sinclair (1991)	Students from kindergarten to grade 10 (N=488)	Qualitative (critical-incident report form)	Five main determinants of students' attitudes towards physical education were found: 1) Curriculum 2) Teacher behavior 3) Class atmosphere 4) Student self-perception 5) Facilities
Mohammed & Mohammad (2012)	Middle and high school students (N=1239)	Quantitative (survey)	1) Students agree that health education should be taught through physical education. 2) Students believe in the importance of physical education classes and agree that physical education grades should be added to the overall grades. 3) Students indicated that physical education classes are fun, make them feel happy, and satisfied. 4) Students acknowledged that physical education classes keep them fit and healthy. 5) Students indicated that they acquire more friends through physical education.
Orunaboka (2011)	Secondary school students (N=112)	Quantitative (questionnaire)	1) Most students exhibited negative attitudes towards physical education. 2) Students' attitudes were positively related to their achievement in physical education.
Phillips & Silverman (2012)	Elementary school students (N=1344)	Quantitative (questionnaire)	The study yielded a valid and reliable instrument for assessing elementary school students' attitudes towards physical education. The developed instrument fit two different models: 1) The first model shows students' attitudes to be influenced by cognition and affect. 2) The second model shows a 4-factor model comprised of cognition-teacher, cognition-curriculum, affect-teacher, and affect-curriculum.

Study	Participants	Method	Findings
Rikard & Banville (2006)	High school students (questionnaire: N=515; interviews: N=159)	Mixed method (survey, interviews)	1) Students preferred a wider variety in sport and fitness activities, an increase in level of challenge, and an increase in student motivation for participating in activities outside school. 2) Student attitudes were accepting or tolerant for participating in fitness activities due to known health benefits. 3) Most students liked physical education class that included some form of game play.
Ryan et al. (2003)	Middle school students (N=611)	Quantitative (survey)	1) The majority of students enjoys having a variety of activities, like the teacher, and had fun in their physical education class. 2) Gender and race variables indicated associations in response to student likes and dislikes of physical education teachers and classes.
Silverman & Subramaniam (1999)		Review	1) Different methods have been used to measure student attitude in physical education, lacking detail about reliability and validity. 2) There is little programmatic research in this area that has produced mixed findings. 3) Future research on student attitude towards physical education needs to investigate attitude from a multidimensional perspective.
Stelzer et al. (2004)	High school students (N=1107)	Quantitative (survey)	1) Students from Czech Republic had significantly higher attitude towards physical education, compared to students from Austria, England, and USA. 2) Males showed more favorable attitude towards physical education than females. 3) Several notable differences were found for the combined effect of gender and country.
Subramaniam & Silverman (2000)	Middle school students (elicitation: N=100; reliability/ validity: N=995)	Quantitative (questionnaire)	Results indicated that the developed instrument for assessing students' attitudes towards physical education produces reliable and valid scores based on the two-component view of attitude.
Subramaniam & Silverman (2002)	Middle school students (N=995)	Quantitative (questionnaire)	Enjoyment and perceived usefulness of the curriculum, as well as a sense of belongingness differentiated high- and low-attitude students.
Subramaniam & Silverman (2007)	Middle school students (N=995)	Quantitative (questionnaire)	1) Overall, students had moderately positive attitudes towards physical education. 2) Higher grades declined in attitude scores.

Tannehill et al. (1994)	High school students (N=314)	Quantitative (survey)	1) Students reported liking physical education because it was coeducational and provided a wide variety of activities. 2) One third of the students indicated that physical education was important/very important. 3) Fun and enjoyment, and teamwork were most valued by the students.
Tannehill & Zakrajsek (1993)	Middle and high school students (N=366)	Quantitative (survey)	1) Students indicated that physical education should teach fitness, sport skills, team and individual sports, and recreational games. 2) Several ethnical differences could be found in questionnaire scores.
Valdez (1997)	Middle school students (N=207)	Quantitative (survey)	1) Students' and parents' attitudes were significantly different. 2) No significant differences were found regarding gender, ethnicity, and socio-economic status in students' attitudes toward physical education.
Zeng et al. (2011)	High school students (N=1317)	Quantitative (questionnaire)	1) Overall mean scores indicated positive attitudes towards physical education in the students. 2) Scores showed significant differences with respect to gender, ethnic group, and socio-economic status.

The outcomes of the various field studies outlined above have remarkable implications for the present study. With the exception of age, almost all the independent variables that have been said to influence students' attitudes toward physical education are implemented in the present study.[35]

2.4 Research Instruments to Measure Attitudes

Attitudes and perceptions have been subject to a lot of research in the last decades, not only in social psychology, but also in other academic fields. As outlined in the previous subchapter, sport science and – to be more precise – sports pedagogy have intensely examined this field. All field research conducted needs instruments to measure the

---

[35] See chapter 3 for a detailed discussion of the instruments and methods used in this field study.

examined subject, either qualitative or quantitative. Lots of constructs for assessing attitude in a purely psychological and thus general way have been designed. Since the present study is concerned with attitudes toward physical education, however, in this subchapter solely instruments for assessing attitudes toward physical education will be discussed.[36] Therefore, instruments developed for assessing attitudes towards physical activity and exercise will not be covered. As in other fields of study, instruments have to be distinguished with reference to whether they are qualitative or quantitative. The distinctive features have been outlined by various authors (cf. e.g., Cohen, Manion, & Morrison, 2011; Lamnek, 2005; Thomas, Nelson, & Silverman, 2010). Since the distinction applies to all fields of research, an explicit discussion is refrained from at this point.

2.4.1 Quantitative Instruments for Assessing Attitudes toward Physical Education

Within the present field of interest, many instruments to assess attitude toward physical education in a quantitative way have been developed, as the review of the current state of research in chapter 2.3.3 has shown. Even though all the above outlined investigations may provide interesting results, most of the papers do not state reliability or validity information (cf. Silverman & Subramaniam, 1999). Beside objectivity, reliability and validity, however, are the central quality criteria of quantitative research (cf. Balnaves & Caputi, 2001; Kromrey, 2006). Silverman and Subramaniam (1999, p. 105) expose the importance of these criteria:

"[...] information on the reliability and validity of an instrument is not provided, the worth of the data may be questioned [...] [because the

---

[36] Silverman and Subramaniam (1999) present a detailed discourse of instruments for assessing attitudes in the above outlined 'general' way.

instruments at hand] may not be measuring the attribute (attitude) they were designed to measure."

As this statement reveals, the collected data needs to be handled with care. This does not mean, however, that such data may not be valuable or interesting. Beside this fact, instruments to measure attitude are always designed to measure the underlying definition of attitude.

As outlined in chapter 2.1.1, attitude has been defined as being uni-dimensional, two-dimensional or multi-dimensional. This, of course, has an immense impact on the construction of an adequate instrument. Most research in physical education – as well as in social science – has complied with the uni-dimensional construction of attitude (cf. Subramaniam & Silverman, 2000). Yet, as outlined in chapter 2.1.1, attitude is currently said to consist of more than one dimension. This implies that most existing research on attitudes toward physical education using the uni-dimensional construct does only investigate one part of attitude (cf. Oppenheim, 2000). Many of the studies illustrated above used either the ATPA instrument developed by Kenyon (1968) or the CATPA instrument designed by Simon and Smoll (1974) to assess attitudes toward physical education.[37]

Even though high internal consistency for the two constructs has been reported (cf. Hagger, Cale, & Almond, 1997), Subramaniam and Silverman (2000, p. 31) criticize the usage of these instruments, since "attitude is not conceptualized as the primary construct". Following the two-dimensional conceptualization of attitude, Subramaniam and Silverman (2000) developed an instrument measuring both the affective and the cognitive dimension of attitude. This instrument only involves the factors of physical

---

[37] Whereas ATPA is the abbreviation for 'Attitude toward physical activity', CATPA refers to the addendum 'children' to design a construct for young people. By implication the CATPA is a modification of the ATPA.

education class content and teacher to measure students' attitudes toward physical education. Albeit, other factors may influence students' attitudes toward physical education as well. The widely used questionnaire developed by Subramaniam and Silverman (2000) is shown in Table 1. It may illustrate typical item phrasings in student attitude research towards physical education.

In the meantime, the Physical Education Activity Attitude Scale (PEAAS) has been prominent in several studies as well (Chung & Phillips, 2002, Park, 1995; Valdez, 1997). In a more recent study, Zeng, Hipscher, and Leung (2011) used the PEAAS to classify pooled student attitude profiles. The questionnaire with a 5-point Likert-type scale with responses ranging from 1 (strongly agree) to 5 (strongly disagree) summed across 20 items. Response values were pooled into a sum score for each participant, resulting in a range of sum scores from 20-100. A score of 20 indicated the most negative attitude, 21-40 indicated a negative attitude, 41-60 indicated a neutral attitude, 61-80 indicated a positive attitude and 81-100 indicated a highly positive attitude.

Ilker and Dermihan (2012) also developed an own instrument that includes attitudes towards physical education, sport and exercise: ATPESS (Attitudes Towards Physical Education and Sport Scale). The questionnaire includes 24 5-point Likert-type items and was validated using a Turkish high school student sample (N=1604). The ATPESS was also used by Arabaci (2009), Chatterjee (2013), and Chatterjee, Nandy, and Adhikari (2012) respectively.

In a recent study, Phillips and Silverman (2012) developed an age group specific instrument for assessing attitudes towards physical education. In a dual-component approach to attitudes that fits with the *Theory of Reasoned Action*, they created a validated instrument for upper elementary school students (fourth and fifth grade).

Table 2: Instrument for assessing students' attitudes towards physical education (Subramaniam & Silverman, 2000)

Item Number	Phrase	Dimension
1.	The games I learn in physical education make my physical education class interesting for me.	Enjoyment (affective component)
2.	The games I learn in my physical education class make learning unpleasant for me.	
3.	The games I learn in my physical education class get me excited about physical education.	
5.	I feel the games I learn in physical education make my physical education class boring for me.	
9.	My physical education teacher makes my physical education class interesting for me.	
11.	I feel my physical education teacher makes learning in my physical education class fun for me.	
12.	I feel my physical education teacher makes my physical education class boring for me.	
15.	My physical education teacher makes learning in my physical education class unpleasant for me.	
19.	My physical education teacher gets me excited about physical education	
20.	I feel the games I learn in my physical education class make learning fun for me.	
4.	My physical education teacher makes my physical education class seem unimportant to me.	Usefulness (cognitive component)
6.	I feel the games I learn in my physical education class are useless to me.	
7.	The games I learn in my physical education class seem important to me.	
8.	My physical education teacher makes my physical education class seem important to me.	
10.	The games I learn in my physical education class are useful to me.	
13.	I feel the games I learn in my physical education class are valuable to me	
14.	The games I learn in my physical education class seem unimportant to me.	
16.	My physical education teacher makes my physical education class useful for me.	
17.	I feel my physical education teacher makes learning in my physical education class valuable for me.	
18.	I feel my physical education teacher makes learning in my physical education class useless for me.	

2.4.2 Qualitative Instruments for Assessing Attitudes toward Physical Education

Despite the fact that most research in the field of physical education – including most of the studies expounded above – has used quantitative instruments to measure attitudes, some investigations have made use of qualitative methods, using interviews in the majority of cases. One issue employing this method is the dependence on the students' ability to verbalize their experiences (cf. Silverman & Subramaniam, 1999). Thus this method may not be adequate in depicting attitudes, at least when investigating children. Another qualitative instrument for measuring students' attitudes is the critical incident form, which has been used in field research and which has been reported to be a useful instrument (cf. Luke & Sinclair, 1991). This instrument constructed by Flanagan (1954) is an open questionnaire which enables participants "to comment freely on selected events in their physical education experience" (Luke & Sinclair, 1991, p. 33). Thus participants can decide for themselves which factors they consider of importance. This procedure may provide substantial information, since these information constitute a first-hand account toward attitudes. Yet, Silverman and Subramaniam (1999) pose the issue of time and changing attitudes regarding this instrument. As the critical incident form is a record of experiences gained in the past, e.g., in elementary school (cf. Luke & Sinclair, 1991), attitude changes may blur the experiences and the actual attitudes at that time (cf. Silverman & Subramaniam, 1999).

As shown in this subchapter, many different approaches to measure attitudes toward physical education have been introduced. The most important issues in choosing an adequate instrument are the method – that is quantitative or qualitative – and the conceptualization of attitude. A field study therefore needs to take these preceding instruments and results into account and decide for the appropriate instrument, perhaps even mix several approaches to provide a meaningful and valid construct.[38]

---

[38] The instrument used in the present study will be discussed in detail in chapter 3.

3 Study Design

In this chapter, the basis for the empirical study will be discussed. Besides the underlying formulation of questions and the resulting hypothesis, this includes the structure of the questionnaire, the procedure to obtain data, the structure of the sample and the statistical procedures undertaken to analyze the data received. One thing has to be remarked in advance: one major concern of the study was to compare the attitudes of students from different types of schools. Thus, the 9th grade was chosen as target group. A detailed discussion of the sample will be given in this chapter.

3.1 Questions and Hypotheses

As already stated in the introduction, the interest of the study was to investigate how students perceive and evaluate their physical education classes. The preceding chapter has shown the severe lack – especially in German research – concerning this important field. Students are the main agents and thus shape those classes, which is why the study focuses on their points of view and underlying differences and correlations.

From this statement, two main goals of the study can be derived:

1. To display the general attitudes of all the students involved in the study toward physical education classes, and to conclude from these to an overall population.

2. To investigate the dependence of those attitudes as dependent variable on socioeconomic factors and physical activity levels.

   – Whether the standard socioeconomic variables gender, type of school, socioeconomic status, and cultural background have an influence on attitudes of students.

   – Whether body specific variables have an influence on students' attitudes.

– Whether grades have an influence on students' attitudes toward physical education classes.

– Whether the physical activity of students themselves, their parents and their peers influence their attitudes.

These main goals of the present study form the very basis of the formation of any questions and the further employment and incorporation of methods for the investigation. Thus, from these goals, the following specific questions can be phrased:

Question 1:

How do students perceive and evaluate their physical education classes?

Question 2:

Does gender influence students' perception and attitude toward physical education classes?

Question 3:

Does the type of school influence students' perception and attitude toward physical education classes?

_____

Question 4:

Does the grade point average influence students' perception and attitude toward physical education classes?

Question 5:

Does the physical education class grade influence students' perceptions and attitude toward the subject?

Question 6:

Does a migration background influence students' perception and attitude toward physical education classes?

Question 7:

Does the socioeconomic status of the students influence students' perception and attitude toward physical education classes?

Question 8:

Does the constitution of students influence their perception and attitude toward physical education classes?

Question 9:

Does the amount of physical activity and exercise of students influence their perception and attitude toward physical education classes?

Question 10:

Does the amount of physical activity and exercise of students' parents influence students' perception and attitude toward physical education classes?

Question 11:

Does the amount of physical activity and exercise of students' peers influence students' perception and attitude toward physical education classes?

Again, these more specific questions formed the guidelines for the methods and the structure of the study, which will be discussed in chapter 3.2. Since some concepts may be misleading, especially the last three questions require an anew reference to chapter 2.1.4, in which physical behavior and exercise are defined.

On the basis of these 11 questions, and on the basis of the current state of research outlined in chapter 2.2, the following hypotheses have been phrased:

Hypothesis 1:

The majority of students express a positive attitude toward physical education classes.

Hypothesis 2:

Male students' attitudes toward physical education classes differ from those of female students.

Hypothesis 3:

Attitudes toward physical education classes differ regarding the type of school students visit.

Hypothesis 4:

Students' grade point averages affect their attitudes toward physical education classes.

Hypothesis 5:

Students' physical education class grades affect their attitudes toward physical education classes.

Hypothesis 6:

Attitudes toward physical education classes of students with a migration background differ from those of students without migration background.

Hypothesis 7:

Students' attitudes toward physical education classes are not dependent on the socioeconomic status.

Hypothesis 8:

Students with a body mass index (BMI) indicating normal weight show more positive attitudes toward physical education classes than those with a BMI indicating under-/ overweight.

Hypothesis 9:

Students who regularly engage in physical activities and exercise show more positive attitudes toward physical education than those, who do not engage regularly in physical activity and exercise.

Hypothesis 10:

Students whose parents regularly engage in physical activities and exercise show more positive attitudes toward physical education.

Hypothesis 11:

Students whose peers regularly engage in physical activities and exercise show more positive attitudes toward physical education.

As already stated above, the different types of hypotheses – variation or correlation – were chosen due to recent findings of research. Moreover, as a standard requirement of empirical research, the above outlined hypotheses have been transformed to corresponding null hypotheses. These concepts have been used for falsification and thus for the acceptance or rejection of the eleven hypotheses above. In this paper, however, the null hypotheses are not enlisted for reasons of clarity. Moreover, the hypotheses will from this point on be abbreviated as H1, H2, … , H11 in the discussion of methods and results.

3.2 Method Outline

This chapter will give a detailed discussion of the methods employed to conduct the data for this study. To be precise, the method, its structure, its contents, the sample, and the process of its generation will be demonstrated. Main goal was to create a construct that was best able to display the complex issue of attitude in the context of physical education classes and to incorporate devices to monitor H1 to H11. As already pointed out in chapter 2.4, the most common instrument to assess attitudes and perceptions in general is a questionnaire, which was the instrument of choice in the present study, as well.

3.2.1 Structure of the Questionnaire

The research was designed as a quantitative study, since – as outlined in chapter 2.3 – basic investigations already exist. The instrument was designed as closed questionnaire, except the questions concerning the socioeconomic status. Since already several questionnaires had been designed before, it seemed practicable and economic to make use of these already existing questionnaires.

The questionnaire designed for the present study contained a total of 67 questions and had three main parts. The first part contained questions concerning the socioeconomic status of the children; the second part was concerned with the students' physical activity and exercise behavior as well as with that of their parents and peers. The third and last part was designed to inquire data to assess students' perceptions and attitudes toward physical education classes.

Socioeconomic Items

The first part of the questionnaire contained 12 questions concerning the socioeconomic status (SES). Besides the standard variable gender, age was requested to sort out any rough derivations from the sample. School type and the name of the school were questioned to lay the groundwork for the examination of H3 and its corresponding null hypothesis. This moreover had the side effect that invalid questionnaires could be sorted out if the school name and the type of school did not match. In addition, height and weight were retrieved to calculate the person's BMI. The BMI was used as a means to operationalize students' constitution. Grade point average, physical education class grade, and citizenship were retrieved to draw any conclusions relating H4, H5, and H6 and the corresponding null hypotheses. Since a migration background is a broad term, it was defined as the students themselves or one of their parents possessing a non-German citizenship.[39] Thus, students with a German citizenship were additionally asked to give the citizenship of their parents. Last, the socioeconomic status was requested by the means of the students' parents' profession.

The parents' profession has been reported to be a reliable marker of the socioeconomic status (cf. Hoffmeyer-Zlotnik & Geis, 2003, p. 126). Therefore, other markers like income and graduation were neglected on the present study, since they have lost their

---

[39] A migration background is a term of much controversy, since it has not undergone a narrow definition. For the purpose of the present study, it has been operationalized as described above. See Koch-Priewe, Niederbacher, Textor, and Zimmermann (2009, p. 31) for a detailed discussion and an approach of a definition of the term.

autonomous significance (cf. Hoffmeyer-Zlotnik & Geis, 2003, p. 126). Thus, the SES was operationalized as a dichotomous variable, with either a good or a bad SES, depending on the parents' occupation.

Physical Activity and Exercise Behavior

The items in the second part of the questionnaire were taken from a study conducted by Pano and Markola (2011), who used six questions to investigate adolescents' extracurricular physical activity and exercise behavior. Five of these questions were employed in the present study. They were, however, slightly modified for the purpose of this study – especially the options regarding the frequency of physical activity and exercise – to permit a more detailed division without any blurring due to a vague scale. Therefore, for the questions concerning exercise and physical activity behavior, a scale with the following options was chosen:

– 5-7 times a week

– 3-4 times a week

– 1-2 times a week

– less than one time a week

The other three questions concerning the reasons for physical activities and exercise, the institutional inclusion, and the reasons preventing students from engaging in more physical activity and exercise were adopted from the original. They served the purpose to investigate the physical activity and exercise behavior more explicitly, which includes potential relations between the items and physical activity and exercise behavior, or regarding the perception and evaluation of physical education classes.

Another reason for including them in the questionnaire was the threat of a potential blurring of physical activity and exercise behavior due to current injuries or diseases.

To include the commonly assumed influence of parents and peers[40] into the study, the same five questions were asked for their parents and their peers, respectively. Thus, potential relations between the behavior of parents and peers and the perceptions and attitudes of students toward physical education classes could be examined.

Perception and Attitude

In the current study, a two-dimensional view of attitude was adopted. Thus, attitude was conceptualized as having a cognitive and an affective dimension. For this reason, students' perceptions and attitudes toward physical education classes were retrieved by the means of an already existing instrument first used by Subramaniam and Silvermann (2000). In their study, attitude is conceptualized two-dimensionally, resulting in an instrument measuring students' attitudes in terms of enjoyment (affective dimension) and usefulness (cognitive dimension). The Cronbach's alpha reliability coefficient was stated to be .92 for the complete instrument (cf. Subramaniam & Silvermann, 2000, p. 37). This instrument was adopted without any changes in items or order of items.

However, the instrument used by Subramaniam and Silvermann (2000) does only include the teacher and the curriculum as indicators for students' attitudes and perceptions toward physical education. Another study by Luke and Sinclair (1991) suggests that besides the key factors curriculum and teacher, other variables influence students' attitudes and perceptions toward physical education classes. In the study, an open-ended questionnaire was used to allow students to comment freely on their

---

[40] The influence of parents and peers as so called agents of socialization is commonly accepted in sport sociology, however, a detailed discussion is neglected here. See Lange (2005, p. 28) for more information.

experiences. Thus, four major determinants of students' attitudes could be worked out. Besides the already mentioned determinants teacher and curriculum – which have been included in Subramaniam and Silvermann's work, as well – the atmosphere and interaction, and the students' self-perception and experience were reported to be major determinants of students' attitudes (cf. Luke & Sinclair, 1991, pp. 41ff). Hence, those two concepts were included in the questionnaire of the present study. This was affected by the means of the two-dimensional view of attitude following the example set by Subramaniam and Silvermann. Consequently, the 20 existing questions regarding teacher and curriculum were altered to fit the two other determinants of attitude, namely interaction and experience. Again, the original order of items was maintained.

Result of this procedure was a questionnaire examining attitude and perception of students toward physical education classes consisting of 40 items. The teacher, the curriculum, the interaction and the students' experiences were chosen as determinants of attitude, providing ten items each. Moreover, each single determinant was viewed as being conceptualized two-dimensional, with an affective and a cognitive dimension, providing five items each for the ten items of every single determinant.

3.2.2   Pretest

Prior to using the above-described construct for the current study, a pretest was implemented to detect incomprehensibilities, inaccuracies, and other forms of errors. A group of 30 9th-grade students was therefore sent the questionnaire with the request to comment on difficulties they had filling out the questionnaire. This made sense, because the final questionnaire had to be designed for 9th-grade students, but was designed by academics. Thus, some terms were taken for granted by the designing group, but could not be understood by the recipients. The pretest thus served as a correction by the means of simplification for the target group to understand all the items. Besides, the 30 students had to take the time they needed for filling out the questionnaire. This average time was needed to inform school administration about the time required for the study.[41] Figure 10 gives a review of the results of the pretest.

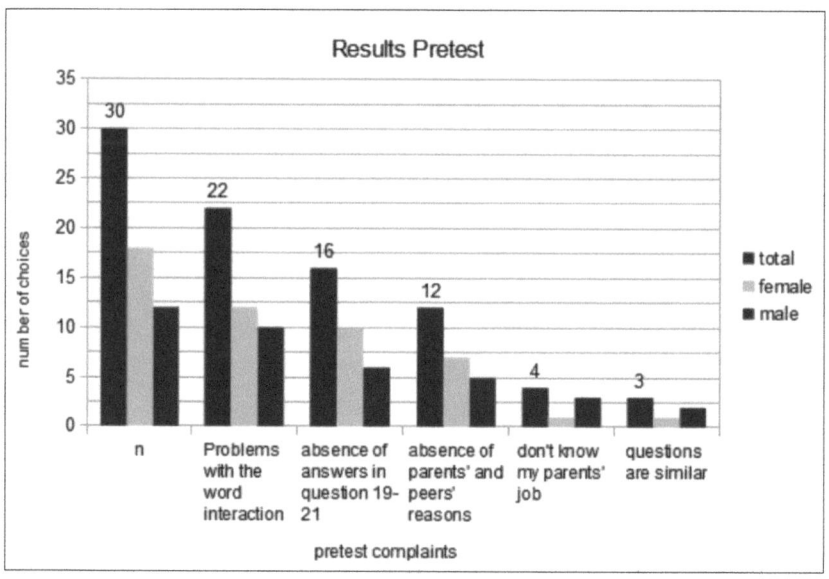

Figure 8: Results of the Pretest, displaying Complaints of the Pretest Group and corresponding Numbers of Choices, for the Whole Group and Separated according to Gender.

The major concern of the pretest group was the term *interaction* originally used in the questionnaire for questions such as *The interaction with my peers make physical education classes important for me*. A huge majority of the pretest group declared that

---

[41] Chapter 3.2.3 deals with the act of execution in more detail.

they did not know the meaning of the word *interaction*. Thus, it was substituted by the term *cooperation and contact*.[42]

Another remark was the absence of certain possible answers in the second part of the questionnaire. The questions about the institutional framework of students' exercise lacked the possible answer *none of them*. Obviously, this was easily corrected by adding such an answer.

The last major comment of the pretest group again concerned part two of the questionnaire. The pretest questionnaire included the questions concerning the reasons to engage in exercise and physical activity and the reasons preventing from engaging more in exercise and physical activity concerning students, but the equivalents for parents and peers were left out. Almost 40 % of the pretest group reported these categories to be lacking. Main explanation was that injuries and diseases prevented parents and peers from engaging more intensively in exercise and physical activity. Thus, these items were also included for parents and peers.

The two minor complaints referred to the lack of knowledge of the parents' job and the seemingly similar questions. Due to the small number of mentions, however, those complaints were rejected as not problematic.

3.2.3 Procedure of the Study

Once the questionnaire had been designed and the target group had been chosen, schools, which supported the execution of the study had to be selected. Schools had to be within a selected locality to obtain comparable data from a certain catchment area with similar conditions. Once the schools were selected, the school administration was

---

[42] Obviously, the questionnaire was designed in German, thus the German counterparts were *Interaktion* changed for *Zusammenarbeit und Kontakt*.

contacted by phone. Moreover, they were sent a document with the main information about the study. When administrations gave their approval, an appointment was arranged to inform classroom teachers of the 9th grades and to look for possible dates to execute the survey. Furthermore, information sheets were distributed by the classroom teachers in class to inform the parents about the study. Students then had to hand in the signed stubs with remarks if the parents did not allow their children to participate in the survey. As soon as the stubs were handed in, an executing person visited the class to supervise the completion of the questionnaires.

For reasons of practicability and economy, the questionnaire was designed as an online questionnaire students were able to access via internet. Thus, the manual data input and with it a possible source of error was avoided. Besides, this procedure saved a lot of time and paper and was therefore both economically and ecologically valuable. To avoid doubly filled in questionnaires by the same person, the questionnaire was filled in during a lesson in the local computer lab and supervised by an executing person. The link by which the questionnaire could be accessed was only activated for this period. Thus, the possibility for a manipulation was kept to a minimum. Besides, this had the additional effect of getting a maximum of students to answer the questionnaire. Students were guided through the online questionnaire by guidelines before each of the three parts that should help them to fill in the questionnaire correctly. Hence, again, the potential for misunderstanding and – as a result – invalid data was minimized.

During the whole process of establishing contact with school administrations, teachers, and parents, the logo of both the University Stuttgart, Germany, and of the Department of Sport and Exercise Science were used. Likewise, the questionnaire was equipped with those logos, respectively.

3.2.4 Interpretation of Variables

As shown above, many variables were used and thus needed to be operationalized. This subchapter will therefore deal with the operationalization of the variables used in the different parts of the questionnaire. The variables of the first part were, if possible, operationalized according to their nature.

- Gender and citizenship were nominally scaled and interpreted due to their nature as dichotomous variable, i.e. gender as male or female and citizenship as German or migration background.

- The type of school was ordinally scaled, with the Gymnasium indicating the highest and the Haupt-/Werkrealschule indicating the lowest form.

- Students' height and weight were used to calculate the BMI. Interpretation of the BMI followed from the classification of the BMI as described in World Health Organization (2000, p. 8). According to this, a BMI between 15.00 and 18.49 is classified as underweight, a BMI between 18.5 and 24.99 is classified as normal weight and a BMI between 25.00 and 29.99 is classified as overweight. Moreover, a BMI below 15.00 is classified as anorexic and a BMI above 30.00 is classified as obese. Thus, in the present study the variable was classified as tripartite: normal weight, under- or overweight, and anorexic/obese.

Figure 11 displays the different categories. On the basis of this classification, the BMI was operationalized as ordinally scaled in the present study, with normal weight as the 'good' form ranging to anorexic/obese as the 'bad' form.

Classification	BMI	Risk of comorbidities
Underweight	<18.50	Low (but risk of other clinical problems increased)
Normal range	18.50–24.99	Average
Overweight:	≥25.00	
Preobese	25.00–29.99	Increased
Obese class I	30.00–34.99	Moderate
Obese class II	35.00–39.99	Severe
Obese class III	≥40.00	Very severe

Figure 9: A classification of the BMI (Source: World Health Organization, 2000, p. 9).

- Grades of physical education class were interpreted on an ordinary scale, and grade point averages were interpreted on an interval scale according to their nature.

- The SES was ordinally scaled. Therefore, a simple division between good and bad SES was made. Good SES and a bad SES were distinguished in a simple way: students were asked to give their parents' jobs. Hoffmeyer-Zlotnik and Geis (2003, p. 126) emphasize the importance of the parents' jobs for the SES. Therefore, the SES in the present study was assigned as follows. If both parents worked (no temporary jobs) or if at least one parent had a job indicative for an academic graduation, a good SES was assigned. In all other cases, a bad SES was assigned.

The variables of the second part of the questionnaire were interpreted in a similar manner:

- The first three questions concerning the exercise frequency of students and their parents and peers was ordinally scaled on a scale ranging from *5-7 times a week* to *less than once a week*.[43]

- In the same way, questions 4 to 6 of the second part of the questionnaire about the frequency of physical activity of students and their parents and peers were ordinally scaled.

- Questions 7 to 15 about the reasons for, the context of, and the reasons preventing from physical activity and exercise were nominally scaled according to their nature. Multiple answers were possible in this section. As already mentioned, those questions were only used to describe the physical activity and exercise behavior in more detail.

The dependent variables of attitudes and perceptions of students toward physical education classes provided the key element of the questionnaire and the study. As already shown above, 40 questions were asked regarding the curriculum, the teacher, the interaction, and the experience, providing 10 questions each. Each of those four factors consisted again of two subfactors, namely the *enjoyment* and the *perceived usefulness* subfactors, displaying the affective and the cognitive dimension, respectively[44]. Answers to the questions were provided by means of a five-point Likert scale, ranging from *strongly disagree* over *disagree, neutral,* and *agree* to *strongly agree*. For the reason of the equivalent distance between each possible answer, those items were analyzed on the basis of an interval scale. Therefore, answers to negative statements such as *my physical education teacher makes my physical education class seem unimportant to me* were reversed during the analysis to display the correct result.

---

[43] See also chapter 3.2.1 for further information about the possibilities for answers.

[44] The terms *enjoyment* and *perceived usefulness* were used by Subramaniam and Silvermann (2000) in the original questionnaire and are therefore adapted here.

Then, the total sum and average score could be calculated for the attitudes as a whole, for all four determining factors, and for the subfactors. These outcomes served as point of reference for the examination of the above-presented hypotheses H1-H11, or to be precise for the attempt of falsification of the corresponding null hypotheses.

3.2.5 Sample

As already indicated at the beginning of this chapter, the study was designed as a comparison between the various types of schools. Obviously, this is reflected in the above stated hypotheses. Moreover, this decision decisively influenced the structure of the sample, since Haupt-/Werkrealschulen end after the 9th grade, Realschulen end after the 10th grade and Gymnasien end after the 12th or 13th grade. Thus, a grade occurring in all three types of schools had to be chosen. Since it is commonly accepted that the cognitive ability develops during adolescence (cf. Mietzel, 2002; Schneider & Büttner, 2002), the highest mutual grade was chosen as target group. Moreover, schools were selected with regard to a defined catchment area to avoid different prerequisites.

Thus, a total sample of 322 students was drawn. This sample contained 158 male and 164 female students. Table 3 displays the distribution regarding gender.

Gender

		Frequency	%	valid %	cumulated %
Valid	female	164	50,9	50,9	50,9
	male	158	49,1	49,1	100,0
	total	322	100,0	100,0	

Table 3: Gender Distribution of the Sample.

Regarding the type of school, 136 students form Gymnasien, 119 students from Realschulen, and 67 students from Haupt-/Werkrealschulen were sampled. Figure 12 and Table 4 illustrate this distribution.

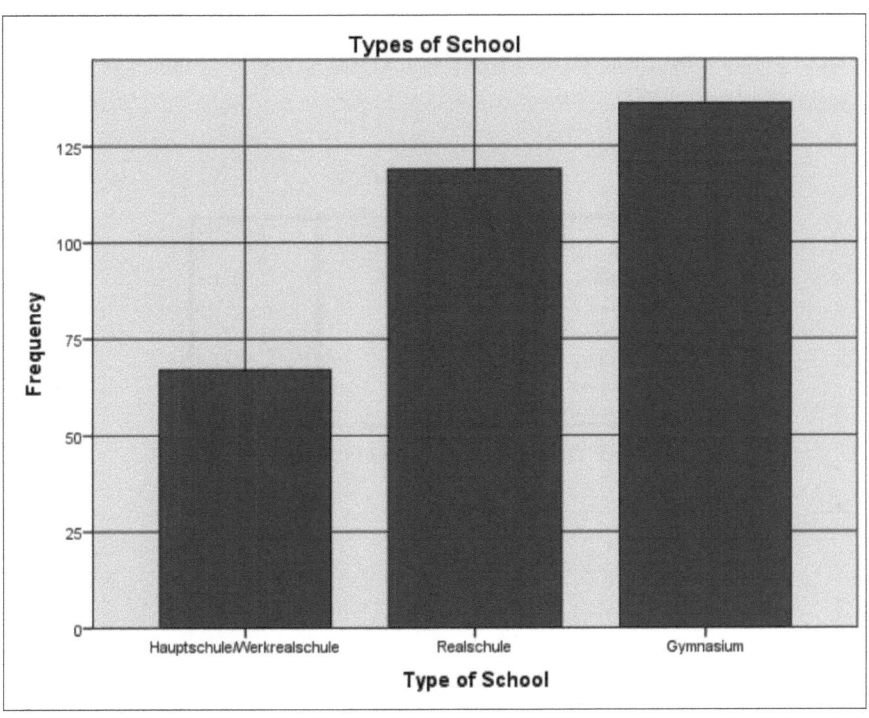

Figure 10: Sample Composition regarding the Type of School.

Type of School

		Frequency	%	valid %	cumulated %
Valid	Hauptschule/Werkrealschule	67	20,8	20,8	20,8
	Realschule	119	37,0	37,0	57,8
	Gymnasium	136	42,2	42,2	100,0
	Total	322	100,0	100,0	

Table 4: Sample Composition regarding the Type of School.

The age of the students who were questioned during the survey was partly determined by the setting of the study: as described above, only the 9th grades of all types of schools were included in the sample. Still, the age may vary within a particular grade level. As table 3 shows, the mode of the sample was 15 years, with a minimum of 13 and a maximum of 17.

Statistics

age

N	Valid	322
	Lacking	0
Median		15,00
Mode		15
Standard Derivation		,763
Range		4
Minimum		13
Maximum		17

Table 5: Descriptive Statistics of the Students' Age.

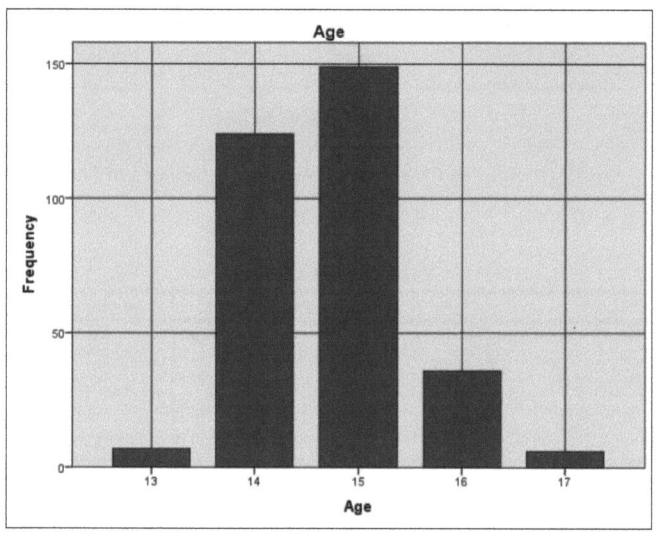

Figure 11: Frequency Distribution of Students' Age.

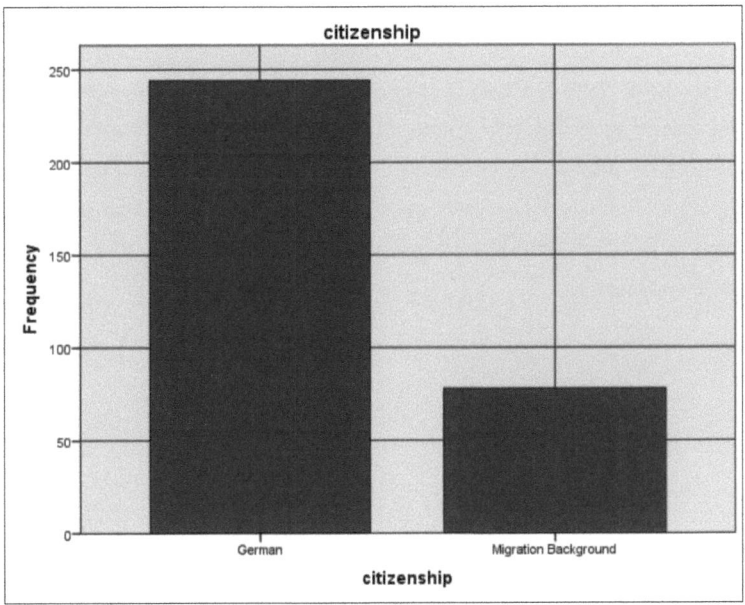

Figure 12: Distinction of the Sample with reference to citizenship.

However, 95.9 % of the questioned students were between 14 and 16. Nevertheless, all answered questionnaires were evaluated, since the reported range seems to display the normal distribution of a 9th grade. Figure 11 displays the frequency distribution of the students' age.

Regarding the citizenship of students, the sample consisted of a majority of students with a German passport and less students with a migration background. However, since whole classes of schools were surveyed, this distribution could not be influenced by the setting. Thus, 244 students with German passports and 78 students with a migration background were included in the study, which is illustrated in Figure 12.

3.3 Statistics

Altogether, three main software programs were used to conduct and evaluate the obtained data. The questionnaire was made accessible via the questionnaire online platform Unipark provided by Questback. The access authorization was provided by the Department of Sport and Exercise Science of the University of Stuttgart, Germany. Moreover, IBM SPSS Statistics 19 and Microsoft Office Excel 2007 were used to analyze the obtained data.

4 Results

In the previous section, the complete structure of the present study was demonstrated. This included the structure of the questionnaire, the structure of the sample, the procedure, and most importantly the underlying hypotheses. In this chapter, the results of the survey will be presented. Therefore, the data will be exposed on a descriptive basis first, and then, in the second part, this data will be analyzed in reference to the eleven hypothesis introduced in the preceding chapter. All tables and figures not included in the main part of this study can be found in Appendix 1.

4.1 Descriptive Statistics

As already described in chapter 3, several independent variables were used to investigate students' attitude toward physical education classes. Among those independent variables were students' gender, physical education grade, grade point average, BMI, SES, type of school, citizenship, and the exercise and physical activity behavior of students, their parents and their peers. Before hypotheses H1 to H11 will be investigated, it is necessary to take closer look at the distributions and frequencies of the above-mentioned items, serving as independent variables.

The distributions of the independent variables gender, type of school, and citizenship have already been described in detail in chapter 3.2.5. The sample consisted of approximately equivalent male and female students, however, those with a migration background were distinctly outnumbered – due to the setting of the study. Likewise, more students attending a Gymnasium were included in the sample.[45] The BMI of students was calculated on the basis of the given height and weight. Figure 15 shows

---

[45] Chapter 5 will discuss and interpret those results in more detail.

that a majority of students had a BMI between 15 and 25. According to the World Health Organization (2000, p. 9), these values are considered to be slightly under- and normal weight.

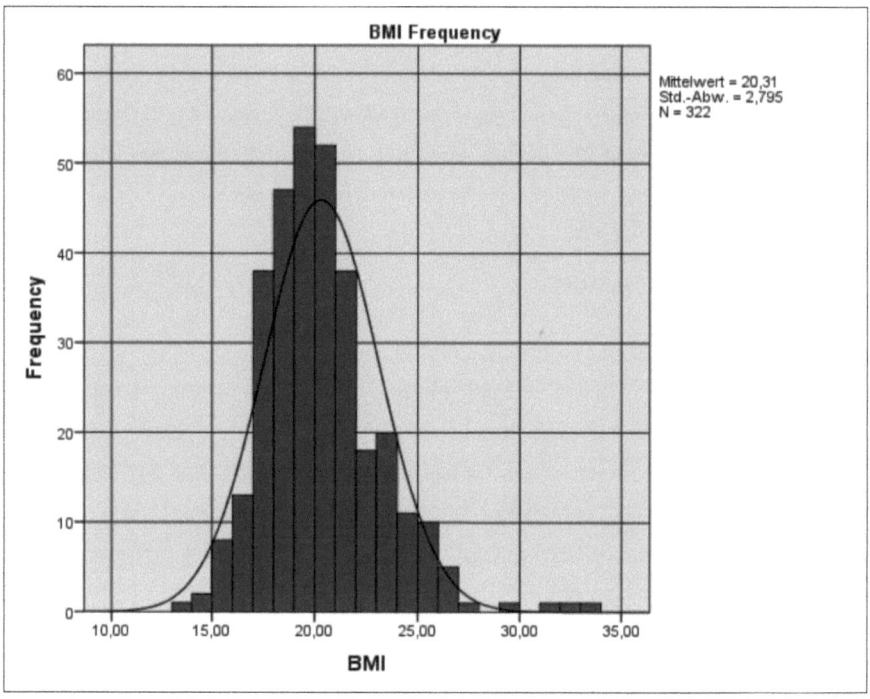

Figure 13: Distribution of Students' BMI.

For the present study, however, students' BMI was classified according to the World Health Organization (2000, p. 9), and therefore all tests were performed with the independent variable of a classified BMI. Students' classified BMIs are displayed in Table 6, showing that 98.1 % of all students were either under-/overweight or normal weight. Only 1.9 % of all students were classified as obese or anorexic and therefore represented the extreme category. The table further shows that 69.5 %, i.e. more than a third of the students were classified as normal weight. On the other hand, this means that 28.7 % of the students were either under- or overweight.

Table 6: Students' BMI as classified according to the WHO (2000, 9).

BMI classified

		Frequency	%	Valid %	Cumulated %
Valid	normal weight	223	69,3	69,5	69,5
	under-/overweight	92	28,6	28,7	98,1
	anorexic/obese	6	1,9	1,9	100,0
	Total	321	99,7	100,0	
Missing	System	1	,3		
Total		322	100,0		

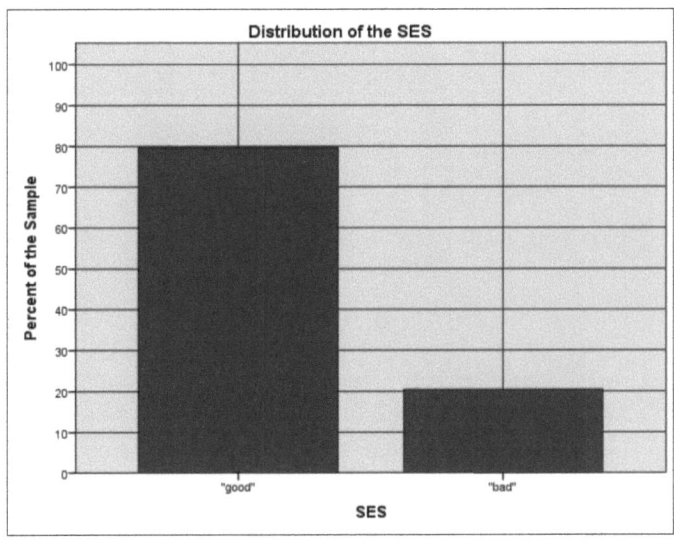

Figure 14: Distribution of Students' SES.

As outlined in chapter 3.2.4, the SES was designed as a dichotomous variable, distinguishing between a good and a bad SES. Interestingly enough, this distinction led to an uneven distribution of the SES within the sample. Whereas only 65 students were ascribed a bad SES, 254 students were ascribed a good SES.

The information of 3 students was not valid. Expressed in percent, 78.9 % percent of the students were classified as having a good socioeconomic status, and 20.2 % as having a bad socioeconomic status. 0.9 % of the students' answers were not valid. This possibility for invalid answers was based on the nature of the SES, which made an open-ended answer necessary, which constituted a field of possible violation. Figure 16 illustrates the distribution of the SES of the present sample.

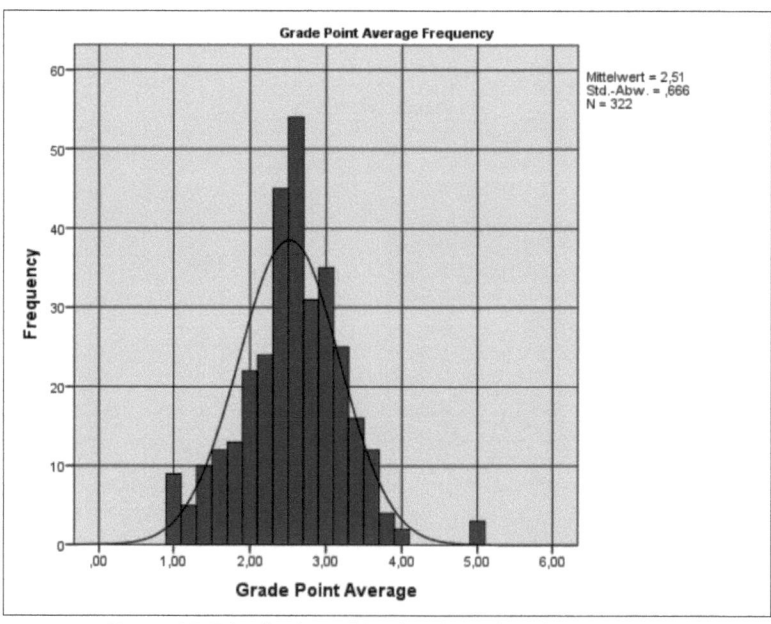

Figure 15: Distribution of Students' Grade Point Averages.

Students' grade point averages were mostly distributed between grade 1 and grade 4, with an increase of frequency around the value of 2.5, a mean of 2.51 and a mode of 2.6. Besides, the median was located at 2.5. The distribution of students' grade point averages is shown in Figure 17.

Beside the by now described variables, the variables physical education grade, the exercise and physical activity behaviors are of particular interest for the present study and the field of research of physical education, since their distributions already display tendencies and provide information about the subject of students in physical education and exercise.[46]

Table 7: Distribution of Students' Physical Education Grades.

Physical Education Grade

		Frequency	%	Valid %	Cumulated %
Valid	1	77	23,9	23,9	23,9
	2	163	50,6	50,6	74,5
	3	71	22,0	22,0	96,6
	4	7	2,2	2,2	98,8
	6	4	1,2	1,2	100,0
	Total	322	100,0	100,0	

Regarding the physical education grades of students, table 5 shows that 96.6 % of all students received a grade between 1 and 3. Grades worse than 3 were not often given. The mode of the sample was 2, likewise the median was 2. Interestingly enough, the number of students – 23.9 % – who received the grade 1 was greater than those who received a 3 (22.0 %).

---

[46] The interpretation of the provided data will follow in chapter 5.

Students' reported exercise and physical activity behavior is illustrated in Tables 7 and 8.

Table 8: Exercise Behavior of Students.

Exercise Frequency

		Frequency	%	Valid %	Cumulated %
Valid	5-7 times per week	43	13,4	13,4	13,4
	3-4 times per week	129	40,1	40,1	53,4
	1-2 times per week	107	33,2	33,2	86,6
	less than 1 time a week	43	13,4	13,4	100,0
	Total	322	100,0	100,0	

Table 9: Physical Activity Behavior of Students

Physical Activity Frequency

		Frequency	%	Valid %	Cumulated %
Valid	5-7 per week	79	24,5	24,5	24,5
	3-4 per week	85	26,4	26,4	50,9
	1-2 per week	102	31,7	31,7	82,6
	less than 1 time a week	56	17,4	17,4	100,0
	Total	322	100,0	100,0	

The tables show that a majority of students exercised between 1-2 and 3-4 times a week, with the mode and the median located at 3-4 times exercise a week. Only 26.6 % of the students reported the extreme forms of *5-7 times a week* and *less than once a week*, each providing 13.3 %. Students' physical activity behavior, on the other hand, was far more balanced. As table 7 shows, the first three categories provided an even distribution of students' physical activity frequency. Compared to the exercise behavior, the mode was located at 3-4 times a week, as well. The median, however, was located at 1-2 times physical activity a week.

Interestingly enough, students reported their peers to have a similar exercise and physical activity behavior. The majority of students reported their peers to exercise between 1-2 and 3-4 times a week, as well. Similarly, mode and median were located at category 2, *3-4 times a week*. Regarding the physical activity behavior, mode and median were both located at *1-2 times a week*. Tables 10 and 11 display the distributions.

Table 10: Exercise Behavior of Students' Peers.

Peers' Exercise Frequency

		Frequency	%	Valid %	Cumulated %
Valid	5-7 times per week	22	6,8	6,8	6,8
	3-4 times per week	150	46,6	46,6	53,4
	1-2 times per week	111	34,5	34,5	87,9
	less than 1 time a week	39	12,1	12,1	100,0
	Total	322	100,0	100,0	

Table 11: Physical Activity Behavior of Students' Peers.

Peers' Physical Activity Frequency

		Frequency	%	Valid %	Cumulated %
Valid	5-7 times per week	58	18,0	18,0	18,0
	3-4 times per week	96	29,8	29,8	47,8
	1-2 times per week	109	33,9	33,9	81,7
	less than 1 time a week	59	18,3	18,3	100,0
	Total	322	100,0	100,0	

In comparison to the students, their parents showed a much distincter exercise and physical activity behavior as reported by the students. Mode and median of parents' exercise behavior were located at category 3, *1-2 times per week*. Even so, 27.3 % of the parents exercised less than once a week, compared with 13.4 % of the students in this category. Tables 12 and 13 illustrate the detailed findings of students' parents' exercise and physical activity behavior.

Table 12: Exercise Frequency of Students' Parents.

Parents' Exercise Behavior

		Frequency	%	Valid %	Cumulated %
Valid	5-7 times per week	22	6,8	6,8	6,8
	3-4 times per week	62	19,3	19,3	26,1
	1-2 times per week	150	46,6	46,6	72,7
	less than 1 time a week	88	27,3	27,3	100,0
	Total	322	100,0	100,0	

Likewise, mode and median of parents' physical activity behavior were located at category 3, with 71.4 % of the parents engaging in physical activity less than 3 times a week, compared to 49.1 % of the students. Thus, on the one hand – on a purely descriptive level – similarities between students and their peers could be found, but on the other hand, discrepancies between students and their parents were displayed in the obtained data.

Table 13: Physical Activity Frequency of Students' Parents.

Parents' Physical Activity Behavior

		Frequency	%	Valid %	Cumulated %
Valid	5-7 times per week	31	9,6	9,6	9,6
	3-4 times per week	61	18,9	18,9	28,6
	1-2 times per week	122	37,9	37,9	66,5
	less than 1 time a week	108	33,5	33,5	100,0
	Total	322	100,0	100,0	

As to the exercise behavior, the obtained data provided a more detailed description of the institutions involved, the reasons for exercise and the reasons preventing them from engaging more in exercise. Appendix 1 lists the detailed tables of answers. Here, only the most important findings will be presented.

Considering the institutions, 58.7 % of the students reported to be a member of a sports club, and only 23.0 % reported not to be a member of any sports institution at all. As to their peers, 71.7 % of the students reported their peers to be members of a sports club, and only 8.4 % not be a member of any institution. On the other hand, only 29.5 % of their parents were stated to be members of sports clubs, with 31.7 % having no membership at all. Interestingly enough, the findings display a rank, with peers on the one end, parents on the other end and students in between.

Regarding the reasons to engage in exercise, students indicated the fun factor (75.2 %), the health benefits (54.7 %), the meeting of friends (53.7 %), and the increase of one's achievement potential (50.0 %) as major reasons. Similarly, for their peers, the fun factor (69.6 %), the meeting of friends (52.2 %), and the health benefits (46.3 %) ranked within the top four. Only the improvement of one's appearance (49.1 %) differentiated the top rankings of students and their peers. In contrast to these two groups, for their parents, the health benefits (60.9 %), the control of their weight (40.7 %), the prevention of aging (36.0 %), and the fun factor (33.2 %) were reported to be the main reasons for exercise. These results seem to display the same tendencies as the exercise and physical activity behavior. Students and their peers seem to have similar reasons for exercising, whereas other reasons are more important for their parents.

With respect to the reasons preventing to engage in more exercise, student particularly mentioned the time factor (73.0 %). 26.1 % answered with *I just don't want to*. Likewise, students reported their peers to have no time (61.5 %) or that they just didn't want to (40.7 %). Almost the same accounts for their parents. The time factor (72.4 %) and the *they just don't want to* factor (23.0 %) were those most frequently mentioned.

All other answers could be neglected. The answers to this question display the tendency that the before mentioned factors seem to be the main reasons preventing all three groups from engaging more in exercise.

Beside the distribution of the independent variables described above, the statistics of students' attitudes and perceptions toward physical education classes are of major interest of this study to provide a basis for the following examination of the eleven hypotheses. Therefore, students' attitudes and perceptions were analyzed according to the structure of the questionnaire as described in chapter 3.2.1. Table 14 provides statistical information about students' attitudes, the curriculum and teacher based attitudes (as used by Subramaniam and Silvermann, 2000), the interaction and experience based attitudes, and the affective and cognitive subfactors.

Table 14: Statistics of the Different Conceptualizations of Attitude.

Attitudes

		Attitude	Attitude_ct	Attitude_ie	Attitude_af	Attitude_cog
N	Valid	322	322	322	322	322
	Missing	0	0	0	0	0
Mean		3,5615	3,4820	3,6410	3,6189	3,5040
Median		3,5000	3,5250	3,6000	3,6000	3,4500
Mode		3,75	3,00ª	3,00	3,00	3,00
Standard Derivation		,68997	,81182	,70695	,72074	,70741
Minimum		1,53	1,20	1,55	1,40	1,40
Maximum		4,98	5,00	5,00	5,00	5,00
Perzentile	25	3,0250	2,9000	3,0500	3,0000	3,0000
	50	3,5000	3,5250	3,6000	3,6000	3,4500
	75	4,1250	4,1000	4,1500	4,2125	4,0500

a. Mehrere Modi vorhanden. Der kleinste Wert wird angezeigt.

As Table 14 shows, the means of the variable attitude and of all the single conceptualizations varies from a maximum of 3.641 to a minimum of 3.482, thus the range of 0.159 displays no huge difference between the different concepts. The same accounts for the median – with a range of 0.15. Considering the standard derivation, the complex whole concept of attitude shows the slightest derivation of 0.68997. For these reasons, the model of attitude as being conceptualized two-dimensionally and consisting of teacher, curriculum, interaction, and experience, was maintained. Figure 18 illustrates the distribution of students' attitudes toward physical education classes as conceptualized in the present study.

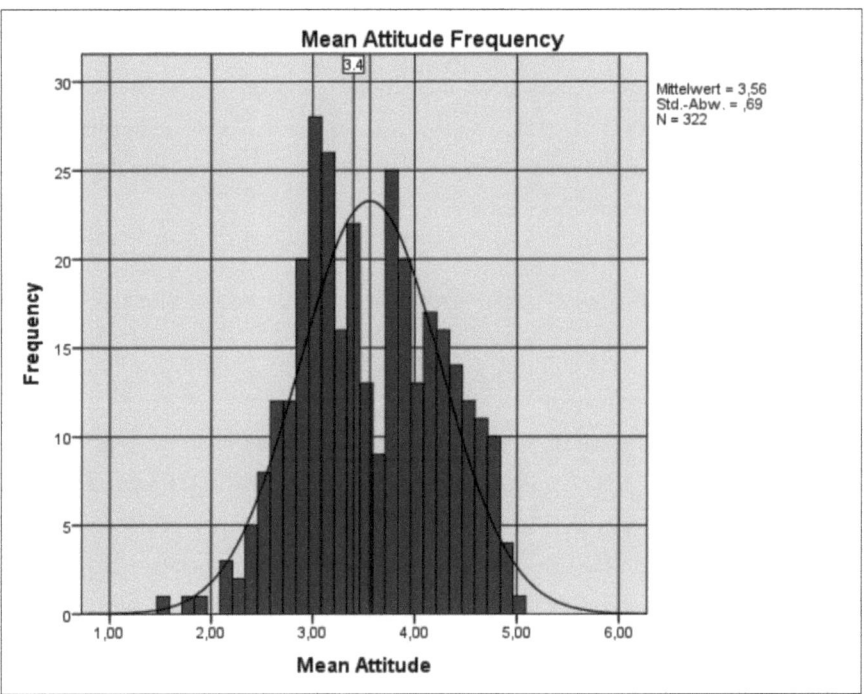

Figure 16: Distribution of Students' Attitudes.

The graph slightly resembles a normal distribution, and a Kolmogorov-Smirnov-Test with a significance level of 0.05 computed via SPSS confirms this assumption, resulting in a significance of 0.09.[47] For this reason, in the following analysis, the dependent variable was considered to be normally distributed, with all its implications for statistical testing.

4.2   Inferential Statistics

As already indicated in the previous chapters, in the present study the attitudes of students toward physical education classes served as dependent variable. All other variables described above served as independent variables to examine the before mentioned hypotheses H1 to H11. Therefore, in this subchapter, those hypotheses will be analyzed, one after the other, on the basis of the descriptive findings demonstrated above. For all tests a significance level of 0.05 was chosen.

H1: The majority of students express a positive attitude toward physical education classes.

Basis for the examination of this hypothesis was the determination of a point of reference, which the mean at hand needed to be compared with. Since 3.0 was the middle of the scale (50.0 %), indicating a totally neutral attitude, a value above this neutral value was chosen as point of reference for the examination of H1. It was set at 60.0 %, which was a mean value of 3.4. Taking the neutral zero point as starting point, the point of reference was 20 % higher than neutral and therefore indicated a positive attitude. The mean of students' attitudes had to be higher than this point of reference. Figure 18 illustrates the location of the mean as the right (non-labeled) line of reference, and the location of the point of reference (left labeled line) on the x-axis. Therefore, on a purely descriptive level, the hypothesis seems to be valid.

---

[47] See Appendix 1 for the corresponding SPSS table.

Table 15: Results of the t-test for Random Samples.

t-test for a Random Sample

	Point of Reference = 3.4				95% Confidence Intervall of Difference	
	T	df	Sig. (2-sided)	Mean Difference	Lower	Higher
Attitude Mean	4,200	321	,000	,16149	,0858	,2371

To conclude from the present sample to a population, an additional t-test for a random sample was computed. Table 13 displays the results of this test. T was reported to be 4.200, and a p-value with a significance of 0.00 suggested a rejection of the null hypothesis in favor of the H1 stated above. Students' attitudes were significantly different (higher) than the chosen point of reference. Hence, a general positive attitude of students toward physical education was accepted.

H2: Male students' attitudes toward physical education classes differ from those of female students.

The distribution of gender has been shown in the description of the sample above. The means for both genders were reported to be 3.4887 for female students, and 3.6370 for male students. Thus, male students seemed to have more positive attitudes toward physical education classes than their female classmates, supporting H2. A t-test for independent samples was calculated to review this difference for significance.

The significance of the Levene-Test of equality of Variances was reported to be 0.201. Therefore, the variances could be assumed to be equal. The corresponding T value was stated to be -1.936, with a p-value of 0.054. Thus, the hypothesis H2 was rejected as not being statistically significant, and the corresponding null hypothesis – as indicated in table 14 – was accepted.

Table 16: t-test investigating the differences regarding gender.

t-test for Independent Samples

	Levene-Test of Equality of Variance		t-test for equality of mean value		
	F	Significance	T	df	Sig. (2-sided)
Variances are equal	1,643	,201	-1,936	320	,054
Variances are not equal			-1,938	319,992	,054

H3: Attitudes toward physical education classes differ regarding the type of school students visit.

As the variable gender, the distributions of the independent variable type of school has already been described above. As Table 17 shows, the means of Realschule and Gymnasium were almost equal, differing only in 0.0219. The mean of Hauptschule, on the other hand, was approximately 0.4 different from those of the other two types of school. Likewise, the Hauptschule involved the lowest minimum and the lowest maximum, whereas especially the maximum of Realschule and Gymnasium only differed in 0.03. Thus, the descriptive data suggested a difference between Hauptschule and the other two types of schools, and a similarity between Realschule and Gymnasium.

Table 17: Descriptive Statistics regarding the Different Types of School.

Descriptives

attitude_mean

	N	Mean	Standard Derivation	Minimum	Maximum
Hauptschule/Werkrealschule	67	3,2466	,57441	1,53	4,53
Realschule	119	3,6559	,66043	1,75	4,95
Gymnasium	136	3,6340	,72600	1,93	4,98
Total	322	3,5615	,68997	1,53	4,98

To examine these suggestions, a one-way ANOVA was computed to inspect possible differences between the groups. Table 16 illustrates the results of this ANOVA.

Table 18: One-way ANOVA examining differences regarding the types of school.

ONEWAY ANOVA

attitude_mean

	Square Sum	df	Mean of Squares	F	Significance
Between Groups	8,417	2	4,209	9,297	,000
Within Groups	144,399	319	,453		
Total	152,816	321			

Table 19: Post-hoc Test regarding the Types of School.

Post-hoc Test

attitude_mean
Scheffé-Prozedur

(I) Type of School	(J) Type of School	Mean Difference (I-J)	Standard Error	Significance
Hauptschule/Werkrealschule	Realschule	-,40924*	,10276	,000
	Gymnasium	-,38737*	,10042	,001
Realschule	Hauptschule/Werkrealschule	,40924*	,10276	,000
	Gymnasium	,02188	,08445	,967
Gymnasium	Hauptschule/Werkrealschule	,38737*	,10042	,001
	Realschule	-,02188	,08445	,967

The one-way ANOVA calculated an F-value of 9.297 and reported a p-value of 0.000, leading to the rejection of the null hypothesis, and the assumption of statistically significant differences between the three groups. To further investigate these differences, or to be more precise the question, between which groups differences existed, a post-hoc test was calculated. The Scheffé-procedure reported significant differences between the means of Hauptschule and Realschule, with a p-value of 0.000, and significant differences between Hauptschule and Gymnasium, with a p-value of 0.001. Between the two types of school Gymnasium and Realschule, a p-value of 0.967 was reported. Thus, no significant difference between those two types existed. Table 17 summarizes the results of the Scheffé-procedure. The results of the Scheffé-procedure therefore confirmed the speculations based on the descriptive data. Thus, H3 was accepted and the corresponding null hypothesis was rejected.

H4: Students' grade point averages affect their attitude toward physical education classes.

Since the formulation of this hypothesis implied a correlation, a Kolmogorov-Smirnov-Test was calculated in advance to test the independent variable for a normal distribution. However, a significance of 0.035 suggested that students' grade point averages were not normally distributed[48], and therefore a Spearman's coefficient was computed to examine the correlation of the two variables grade point average and attitude. This non-parametric correlation test indicated a very weak negative correlation of -0.103, implying that attitudes decreased with higher grade point average values, where higher values signified worse grade point averages. Table 18 summarizes the results of the correlation test.

As the significance of the non-parametric correlation test shows, with a value of 0.065, the p-value was reported to be higher than 0.05. Hence, the observed weak correlation was considered not to be significant and consequently, the null hypothesis was assumed and the hypothesis that students' grade point averages influence their attitudes toward

---

[48] Appendix 2 provides detailed tables and graphs of this test.

physical education classes was rejected. Grade point averages seemed not to have a statistical significant influence on students' attitudes.

Table 20: Spearman's Rho and Significance for Grade Point Average and Attitude.

Correlations

			Grade Point Average	attitude_mean
Spearman-Rho	Grade Point Average	Correlation Coefficient	1,000	-,103
		Sig. (2-sided)	.	,065
		N	322	322
	attitude_mean	Correlation Coefficient	-,103	1,000
		Sig. (2-sided)	,065	.
		N	322	322

H5: Students' physical education class grades affect their attitudes toward physical education classes.

As H4, this hypothesis again implied a correlation between the independent variable physical education class grades and the dependent variable attitudes toward physical education. Since the independent variable was interpreted on an ordinal scale – as already pointed out in chapter 3.2.4 – the Pearson's coefficient was abandoned and a Spearman's correlation coefficient was computed instead. The p-value of this non-parametric test was stated to be 0.000, suggesting that the correlation was statistically significant. Moreover, a medium correlation of -0.392 was reported. Since, again, higher variable values signified worse physical education grades, this medium correlation suggested worse attitudes with increasing grade values, or, decreasing attitudes with decreasing grades. Table 19 summarizes the results of the non-parametric correlation test.

As already indicated the correlation coefficient suggested a negative medium correlation between attitudes and physical education grades.

Table 21: Spearman's Rho and Significance for Physical Education Grade and Attitude.

Correlations

			attitude_mean	Physical Education Grade
Spearman-Rho	attitude_mean	Correlation Coefficient	1,000	-,392**
		Sig. (2-sided)	.	,000
		N	322	322
	Physical Education Grade	Correlation Coefficient	-,392**	1,000
		Sig. (2-sided)	,000	.
		N	322	322

**. Die Korrelation ist auf dem 0,01 Niveau signifikant (zweiseitig).

Figure 17: Simple Linear Regression for the variables Attitude and Physical Education Grade.

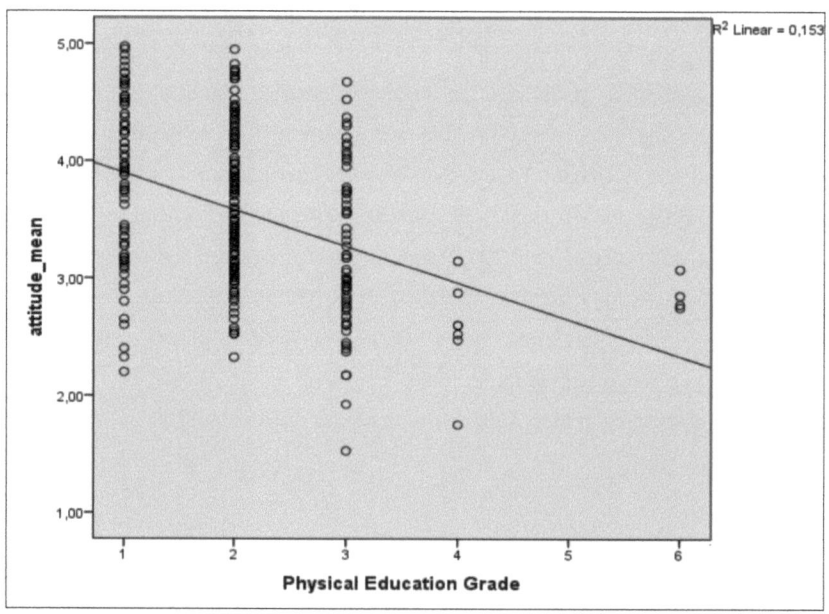

The scatter plot in figure 19 gives the distributions of attitudes separated for physical education grades with the linear regression. This plot served as an orientation concerning the linear regression. As figure 19 shows, r^2 was reported to be 0.153. Due to the very limited expressiveness of the model as reported by the r-/r^2-value, no detailed examinations were undertaken to investigate this correlation any further[49].

H6: Attitudes toward physical education classes of students with a migration background differ from those of students without migration background.

The distributions and the structure of the variable migration background have already been illustrated in chapter 3.2.5. For reasons of simplification, the basic data are restated at this point. The sample consisted of a total of 322 students, of which 78 had a migration background. The rest was categorized as having no migration background.

Figure 20 illustrates the attitudes separated according to the variable migration background. As the diagram suggests, students from the drawn sample without migration background showed better attitudes than their class mates having a migration background. Likewise, the means for the two groups were reported to be 3.6383 for students without migration background and 3.3212 for students with migration background. For the examination of the above stated hypothesis and thus the conclusion to a population, a t-test for independent samples was computed. The Levene-Test of equality of variances stated a significance of 0.086, thus, the variances were assumed to be equal.

---

[49] Nevertheless, Appendix 1 enlists tables concerning the linear regression.

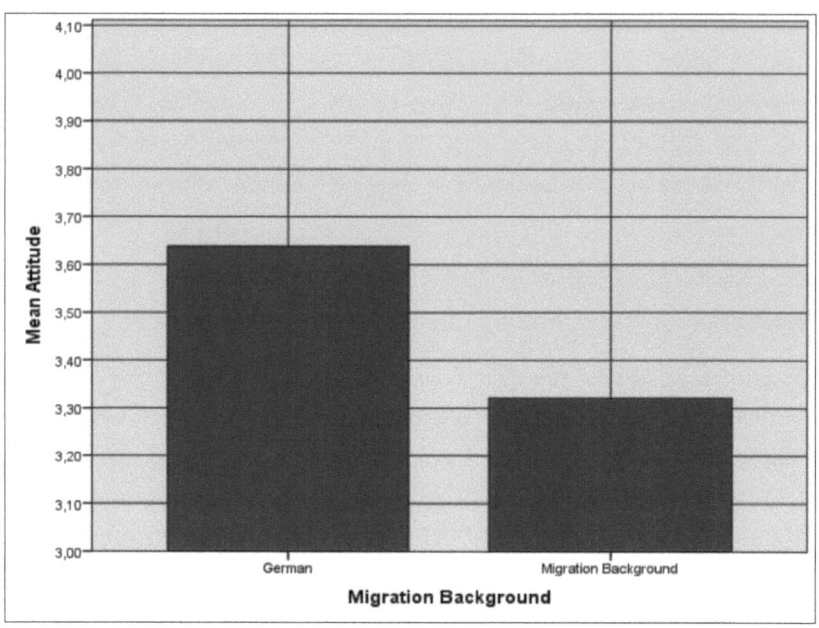

Figure 18: Attitudes Separated according to Migration Background.

Table 22: t-test Investigation of the Differences in Attitude regarding Migration Background.

t-test for Independent Samples

		Levene-Test of Equality of Variances		t-test for Equality of Means		
		F	Significance	T	df	Sig. (2-sided)
attitude_mean	Variances are equal	2,959	,086	3,599	320	,000
	Variances are not equal			3,801	142,782	,000

The t-value for equal variances was calculated as 3.599, and the corresponding p-value was 0.000. Therefore, the difference between the two groups was assumed to be statistically significant, and consequently the null hypothesis was rejected and H6 was accepted. Students without migration background had better attitudes than those with a migration background.

H7: Students' attitudes toward physical education classes are not dependent on the socioeconomic status.

As already outlined in chapter 3.2.4, the SES was designed as a dichotomous variable, distinguishing between either a good or a bad SES. The distributions of good and bad SES has been expounded in chapter 4.1. As to the attitudes of the two groups, students with a good SES showed more positive attitudes than their mates with a bad SES. For the first group, the attitude was reported to be 3.6218, and for students with a bad SES, attitude was stated to be 3.3450. Thus, in the present sample, students of the first group had better attitudes toward physical education classes than students of the second group. Figure 21 illustrates the different means of attitude.

As to H7, a difference considering the attitudes of the two groups seemed to exist in the present sample. Hence, a t-test for independent samples was computed to examine these differences for significance. Even though the hypothesis was formulated "negatively", i.e. that no differences between the two groups existed, the test at hand examined the observed differences between the groups for significance. This implies that the results of the t-test had to be interpreted differently.[50] Table 23 depicts the results of the t-test for independent samples.

---

[50] A p-value below 0.05 would as usually suggest that the differences between the groups were significant. Regarding H7, however, in this case the hypothesis would have to be rejected, since it was formulated "negatively".

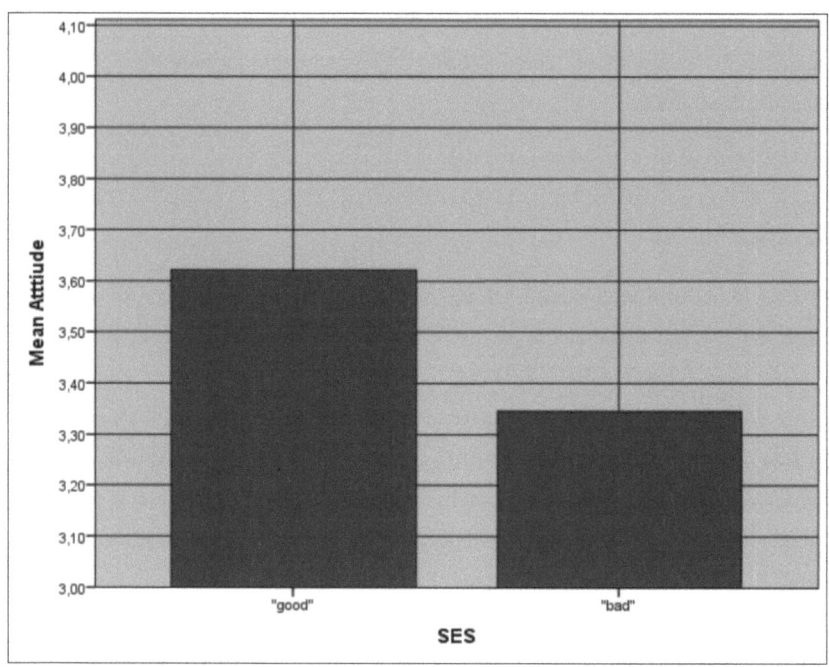

Figure 19: Mean Attitudes for the two SES groups.

Table 23: t-test Investigation of the Differences in Attitude regarding the SES.

t-test for Independent Samples

		Levene-Test for equality of variances		t-test for equality of means		
		F	Significance	T	df	Sig. (2-sided)
attitude_mean	Variances are equal	,725	,395	2,924	317	,004
	Variances are not equal			3,086	106,749	,003

As the results of the Levene-Test for equality of variances indicate, with a significance of 0.395, the variances of the samples could be assumed to be equal. Thus, a t-value of 2.924 was reported, and with it a p-value of 0.004. These results suggested that the differences between the two groups were significant, and therefore applicable to a population. For this reason, H7 – due to its "negative" formulation – had to be rejected and consequently the null hypothesis had to be accepted, which means that the two groups differed concerning their attitudes toward physical education classes.

H8: Students with a BMI indicating normal weight show more positive attitudes toward physical education classes than those with a BMI indicating under/-overweight.

As already pointed out in the previous chapter, the BMI was classified according to the recommendations of the WHO (2000, 9). Considering the data of the current sample, however, the descriptive statistics for the three groups did not prompt more positive attitudes of those students with a BMI classified as normal weight. Normal weight students' attitudes were reported to be 3.5549, under-/overweight students' attitudes were 3.6035, and those of anorexic/obese students were 3.4625. Figure 22 illustrates the attitudes separated for the different classifications of the BMI.

Since the anorexic/obese students constituted only 1.9 % of the total sample, they were ignored for the examination of H8. Thus, a t-test for independent samples could be computed, incorporating only the two groups normal weight and under-/overweight. As the descriptive data shows, under-/overweight showed better attitudes toward physical education classes. Hence, H8 could already be rejected. Nevertheless, the t-test was calculated to examine the different distributions for significance. Table 22 illustrates the results of this test.

The Levene-Test suggested an equality of variances of the two groups, thus the T-value was reported to be -0.575 and the corresponding p-value 0.566. This indicated no statistically significant difference of the means of the two groups.

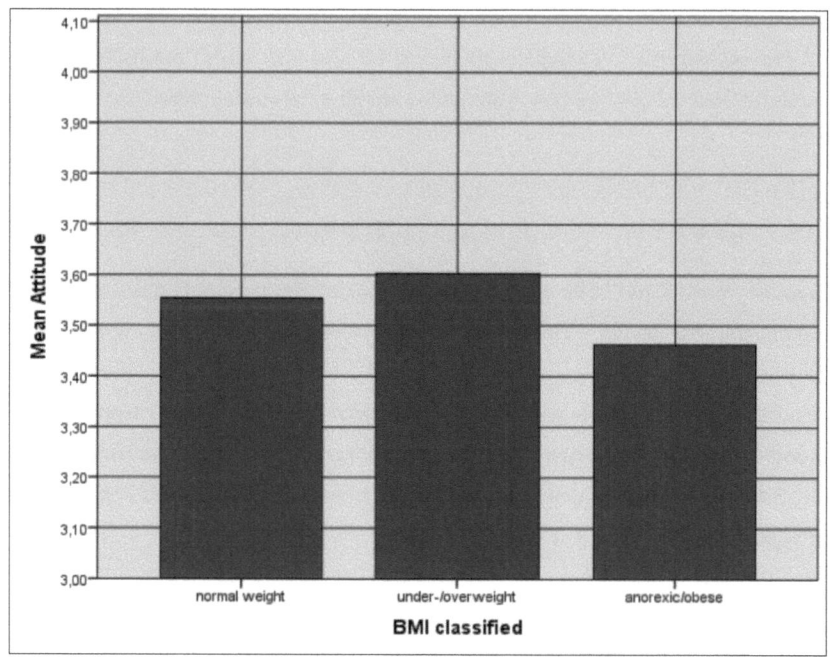

Figure 20: Mean Attitudes for the Three Classified BMI Groups.

Table 24: t-test Investigation of the Differences in Attitude regarding the Classified BMI.

		Levene-Test of Equality of Variances		T-Test for Equality of Means		
		F	Significance	T	df	Sig. (2-sided)
attitude_mean	Variances are equal	1,059	,304	-,575	313	,566
	Variances are not equal			-,557	159,013	,578

H9: Students who regularly engage in physical activities and exercise show more positive attitudes toward physical education than those, who do not engage regularly in physical activity and exercise.

To examine this hypothesis, the two variables exercise behavior and physical activity behavior had to be considered separately. Regarding the exercise behavior of students, mean attitudes in the present sample declined from very regular exercise to irregular exercise, as figure 23 shows. Correspondingly, the means for the different groups were reported to be 3.8215 for those students who exercised 5-7 times a week, 3.6890 for those who exercised 2-4 times a week, 3.5126 for those who exercised 1-2 times a week, and 3.0407 for those who exercised less than once a week. Figure 23 illustrates this distribution. Therefore, on a purely descriptive level, the claim seemed to hold true for the current sample. To check the results for the investigation of H9, a one-way ANOVA was computed to check the distributions for significant differences.

The one-way ANOVA resulted in an F-value of 13.199 and a corresponding p-value 0.000. Hence, a significant difference between – at least – to of the groups was assumed. Subsequently, a post-hoc test was calculated. The Scheffé-procedure was used to determine the groups, between which significant differences existed. Table 23 depicts the results and the data of the post-hoc test. The table shows that the attitudes of those students exercising less than once a week were significantly different from those of the other students. To be precise, for students exercising less than once a week, p-values were 0.000 for the difference concerning students exercising 5-7 times per week, 0.000 for the difference concerning those students exercising 3-4 times per week, and 0.001 for the difference concerning students exercising 1-2 times per week. For the differences between the other groups, no significant p-values were reported. Hence, the first part of H9 was assumed and the corresponding null hypothesis was rejected.

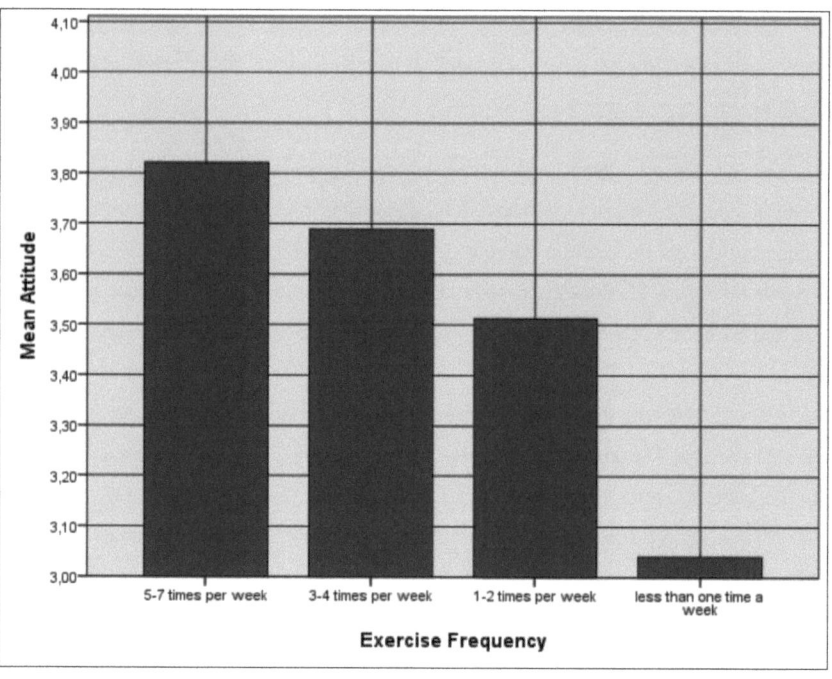

Figure 21: Mean Attitudes Separated according to Exercise Frequency.

However, as stated earlier, H9 consisted of two parts, also involving the physical activity frequency of students. Chapter 4.1 showed that this frequency was distributed more balanced. Figure 24 shows the distributions of the mean attitudes according to physical activity frequency. The means for the four different groups were 3.7816 for students engaging in physical activity 5-7 times a week, 3.5906 for those who engaged in physical activity 3-4 times per week, 3.5637 for those engaging in physical activity 1-2 per week, and 3.2027 for those who engaged less than once a week. Therefore, on a descriptive basis, students of the sample being more physically active reported better attitudes than their more passive classmates.

Table 25: Data of the post-hoc Test for Differences concerning Students' Exercise Behavior.

attitude_mean
Scheffé-Prozedur

(I) Exercise Frequency	(J) Exercise Frequency	Mean Difference (I-J)	Standard Error	Significance
5-7 times per week	3-4 times a week	,13256	,11511	,723
	1-2 times a week	,30889	,11803	,079
	less than 1 time a week	,78081*	,14098	,000
3-4 times per week	5-7 times a week	-,13256	,11511	,723
	1-2 times a week	,17634	,08548	,237
	less than 1 time a week	,64826*	,11511	,000
1-2 times per week	5-7 times a week	-,30889	,11803	,079
	3-4 times a week	-,17634	,08548	,237
	less than 1 time a week	,47192*	,11803	,001
less than 1 time a week	5-7 times a week	-,78081*	,14098	,000
	3-4 times a week	-,64826*	,11511	,000
	1-2 times a week	-,47192*	,11803	,001

*. Die Differenz der Mittelwerte ist auf dem Niveau 0.05 signifikant.

Table 26: Data of the post-hoc Test for Differences concerning Students' Exercise Behavior.

attitude_mean
Scheffé-Prozedur

(I) Frequency of Physical Activity	(J) Frequency of Physical Activity	Mean Difference (I-J)	Standard Error	Significance
5-7 times per week	3-4 times per week	,19106	,10432	,342
	1-2 times per week	,21792	,10005	,194
	less than 1 time a week	,57897*	,11661	,000
3-4 times per week	5-7 times per week	-,19106	,10432	,342
	1-2 times per week	,02686	,09804	,995
	less than 1 time a week	,38791*	,11489	,011
1-2 times per week	5-7 times per week	-,21792	,10005	,194
	3-4 times per week	-,02686	,09804	,995
	less than 1 time a week	,36105*	,11102	,015
less than 1 time a week	5-7 times per week	-,57897*	,11661	,000
	3-4 times per week	-,38791*	,11489	,011
	1-2 times per week	-,36105*	,11102	,015

*. Die Differenz der Mittelwerte ist auf dem Niveau 0.05 signifikant.

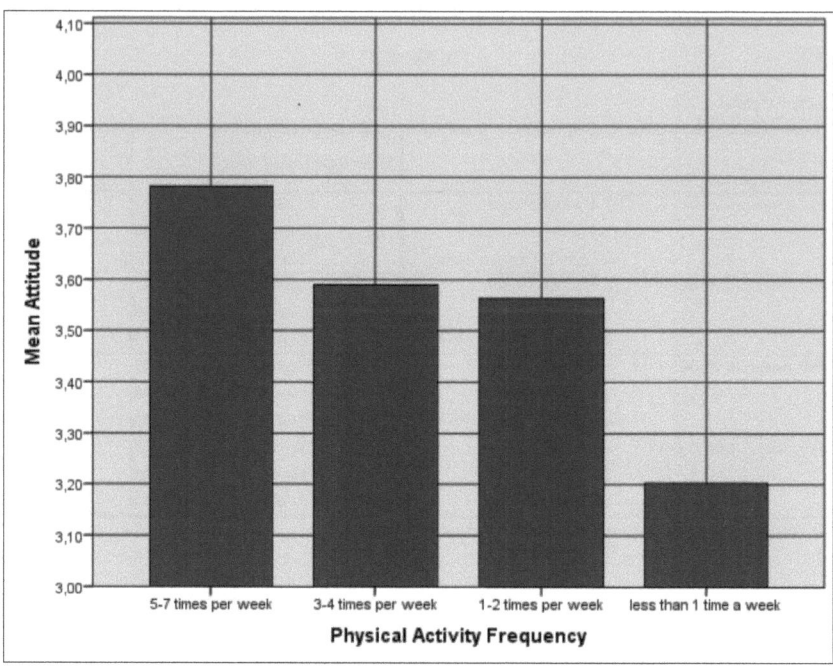

Figure 22: Mean Attitudes Separated according to Exercise Frequency.

As with exercise, the significance of these observed differences was tested with a oneway ANOVA and a succeeding post-hoc test. The F-value of the ANOVA was reported to be 8.312, with a corresponding p-value of 0.000. Hence, a statistically significant difference was assumed between at least two of the four groups. The succeeding post-hoc test resulted in significant differences between the attitudes of those students who were physically active less than once a week and all other groups. For this group, the reported p-values were 0.000 for the difference regarding students engaging in physical activity 5-7 times per week, 0.011 for the difference regarding those being physically active 3-4 times per week, and 0.015 for the difference regarding those students engaging in physical activity 1-2 times per week. For the other groups, no significant p-values were reported. Table 24 summarizes the results of the Scheffé-procedure.

On the basis of this data, the second part of H9 was assumed, as well. Therefore, the whole hypothesis was accepted and the corresponding null hypothesis was rejected.

H10: Students whose parents regularly engage in physical activities and exercise show more positive attitudes toward physical education.

For this hypothesis, the same procedure as for H9 was carried out. Exercise and physical activity of students' parents were considered separately to examine H10. Referring to the amount of students' parents' exercise, the means of attitude showed the following distributions: 3.4648 for students whose parents exercised 5-7 times per week, 3.6520 for those with parents exercising 3-4 times per week, 3.6717 for students with parents exercising 1-2 times per week, and 3.3341 for those whose parents exercised less than once a week, respectively. Again, a oneway ANOVA was computed to test the significance of these differences. The resulting F-value was reported to be 5.154, and the p-value was 0.002, indicating a significant difference between the groups.

The post-hoc test reported significant differences between the group with parents exercising less than once a week and two other groups of students, namely those whose parents exercised 1-2 times a week and those whose parents exercised 3-4 times a week. The p-values for these differences were 0.004 (1-2 times a week) and 0.047 (3-4 times a week), respectively. Interestingly enough, no significant differences between students whose parents exercised 5-7 times a week and those whose parents engaged in sporting activity less than once a week were reported. Still, the first part of H10 was assumed. Table 25 summarizes the data of the post-hoc test.

Table 27: Data of the post-hoc Test for Differences concerning Students' Parents' Exercise Behavior.

attitude_mean
Scheffé-Prozedur

(I) Exercise Frequency of Students	(J) Exercise Frequency of Parents	Mean Difference (I-J)	Standard Error	Significance
5-7 times per week	3-4 times per week	-,18724	,16799	,743
	1-2 times per week	-,20689	,15455	,617
	less than 1 time a week	,13068	,16136	,883
3-4 times per week	5-7 times per week	,18724	,16799	,743
	1-2 times per week	-,01965	,10221	,998
	less than 1 time a week	,31793*	,11225	,047
1-2 times per week	5-7 times per week	,20689	,15455	,617
	3-4 times per week	,01965	,10221	,998
	less than 1 time a week	,33758*	,09090	,004
less than 1 time a week	5-7 times per week	-,13068	,16136	,883
	3-4 times per week	-,31793*	,11225	,047
	1-2 times per week	-,33758*	,09090	,004

*. Die Differenz der Mittelwerte ist auf dem Niveau 0.05 signifikant.

Regarding students' parents physical activity frequency, the means of students' attitude were distributed as follows: 3.6573 for those whose parents engaged in physical activity 5-7 times per week, 3.5414 for students with parents being physically active 3-4 times per week, 3.6166 for students with parents engaging in physical activity 1-2 times per week, and 3.4831 for those whose parents were physically active less than once a week. Figure 25 illustrates this distribution. The diagram shows that students' attitudes toward physical education in the current sample seemed not to follow any pattern. Furthermore, the differences between the four groups seemed to be only marginal.

To examine the differences between the four groups, a one-way ANOVA was computed. The F-value was reported to be 0.940, with a p-value of 0.422. Hence, no statistically significant difference between the groups was assumed. As a result, the second part of and with the whole H10 was rejected, and the corresponding null hypothesis was accepted.

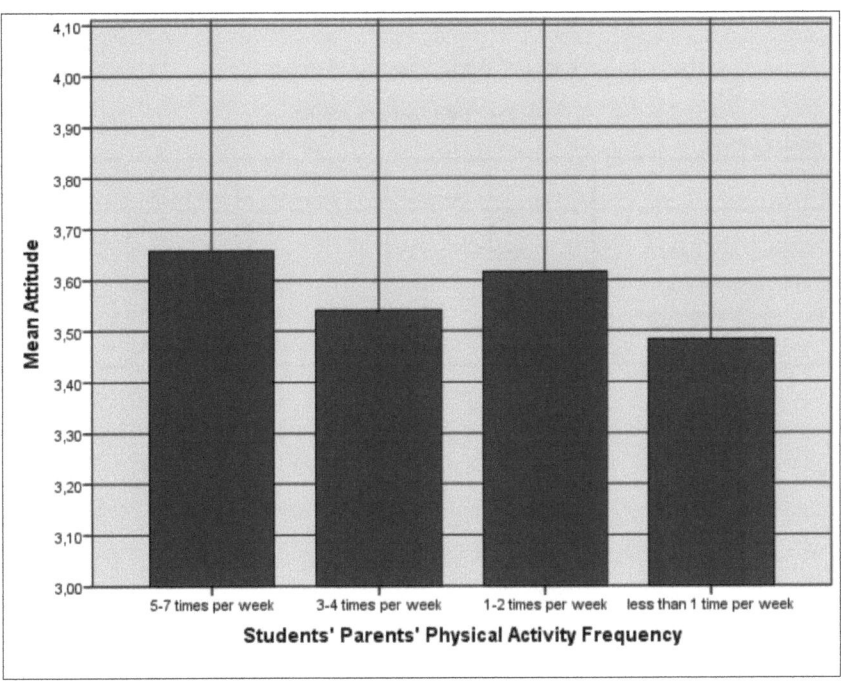

Figure 23: Mean Attitudes Separated according to Students' Parents' Physical Activity Frequency.

H11: Students whose peers regularly engage in physical activities and exercise show more positive attitudes toward physical education.

Similarly to the hypotheses H9 and H10, H11 was divided regarding students' peers' exercise behavior and students' peers' physical activity frequency. Table 26 illustrates the means of students' attitudes for the four different groups.

Table 28: Descriptive Data concerning Students' Attitudes divided according to Students' Peers' Exercise Behavior.

ONEWAY Deskriptive Statistics

attitude_mean

	N	Mean	Standard Derivation	Minimum	Maximum
5-7 times per week	22	3,7182	,69068	2,45	4,83
3-4 times per week	150	3,7302	,66247	1,93	4,95
1-2 times per week	111	3,4462	,68225	1,53	4,75
less than 1 time a week	39	3,1526	,59382	1,75	4,98
Total	322	3,5615	,68997	1,53	4,98

As the table shows, the means for the two highest categories were much higher than those of the two lowest categories. A one-way ANOVA was therefore conducted to examine the differences between the four groups. The resulting F-value of this test was 9.695, and the p-value was 0.000, suggesting a significant difference between the four groups.

The subsequent post-hoc test showed significant differences concerning two groups: first, students' attitudes significantly differed between the group of students whose peers exercised 5-7 times per week and the students whose peers exercised less than once a week (with a p-value of 0.018). Secondly, students whose peers exercised 3-4 times a week showed significantly higher attitudes than those students whose peers exercised less than once a week (0.000), and moreover than those students whose peers exercised 1-2 times per week (0.009). Table 27 illustrates these results of the Scheffé-procedure. As a result, the first part of H11 was assumed to be correct.

The same procedure was carried out for the second part of the hypothesis. The data suggested that for the current sample attitude means for the four groups declined from the most regular to the least regular physical activity level. Table 28 depicts the different attitude means for the four groups.

Table 29: Data of the Post-Hoc Test for Differences concerning Students' Peers' Exercise Behavior.

(I) Exercise Frequeny of Students' Peers	(J) Exercise Frequency of Students' Peers	Mean Difference (I-J)	Standard Error	Significance
5-7 times per week	3-4 times per week	-,01198	,15149	1,000
	1-2 times per week	,27201	,15485	,380
	less than 1 time a week	,56562	,17692	,018
3-4 times per week	5-7 times per week	,01198	,15149	1,000
	1-2 times per week	,28400	,08308	,009
	less than 1 time a week	,57760	,11927	,000
1-2 times per week	5-7 times per week	-,27201	,15485	,380
	3-4 times per week	-,28400	,08308	,009
	less than 1 time a week	,29361	,12351	,132
less than 1 time a week	5-7 times per week	-,56562	,17692	,018
	3-4 times per week	-,57760	,11927	,000
	1-2 times per week	-,29361	,12351	,132

Table 30: Descriptive Data concerning Students' Peers' Physical Activity Behavior.

	N	Mean	Standard Derivation	Minimum	Maximum
5-7 times per week	58	3,7530	,63389	1,75	4,95
3-4 times per week	96	3,6091	,70566	1,93	4,95
1-2 times per week	109	3,5800	,70312	1,53	4,83
less than 1 time a week	59	3,2614	,60911	2,20	4,98
Total	322	3,5615	,68997	1,53	4,98

To investigate the data for significant differences, a one-way ANOVA was calculated, resulting in an F-value of 5.620 and a p-value of 0.001. This suggested a significant difference between the groups. The post-hoc test revealed significant differences only for one group. The attitudes of those students whose peers were physically active less than once a week were significantly lower than the attitudes of all other groups. To be precise, p-values for the differences between the mentioned group and the others were 0.002 (5-7 times per week), 0.023 (3-4 times per week), and 0.038 (1-2 times per week), respectively. The differences between the other groups were reported not to be significant. The results of the post-hoc test are illustrated in Table 31.

Table 31: Data of the post-hoc Test for Differences concerning Students' Peers' Physical Activity Behavior.

attitude_mean
Scheffé-Prozedur

(I) Physical Activity Frequency of Students' Peers	(J) Physical Activity Frequency of Students' Peers	Mean Difference (I-J)	Standard Error	Significance
5-7 times per week	3-4 times per week	,14390	,11235	,651
	1-2 times per week	,17297	,10980	,480
	less than 1 time a week	,49158*	,12491	,002
3-4 times per week	5-7 times per week	-,14390	,11235	,651
	1-2 times per week	,02907	,09455	,992
	less than 1 time a week	,34767*	,11175	,023
1-2 times per week	5-7 times per week	-,17297	,10980	,480
	3-4 times per week	-,02907	,09455	,992
	less than 1 time a week	,31861*	,10919	,038
less than 1 time a week	5-7 times per week	-,49158*	,12491	,002
	3-4 times per week	-,34767*	,11175	,023
	1-2 times per week	-,31861*	,10919	,038

*. Die Differenz der Mittelwerte ist auf dem Niveau 0.05 signifikant.

Since the most irregularly active group showed significantly lower attitudes than the other groups, the second part of H11 was assumed to be correct, as well. Hence in total, H11 was accepted and the corresponding null hypothesis was rejected.

5 Discussion

In the preceding chapters, the basis for an interpretation of the results was formed. Chapter 2 dealt with the current state of research of the field of interest, chapter 3 was concerned with the empirical description of the present study and the results of the present study were described, presented, and analyzed in chapter 4. This section will build on this presentation, discussing the results of the present study with relation to earlier findings of other studies and possible causes for the reported manifestation of the questioned features. Therefore, both descriptive and inferential statistics will be interpreted.

Before emphasis of the discussion is placed on the actual findings of the survey, another important aspect of the study needs to be addressed. With a total of 322 students involved in the survey, the sample is relatively small. This circumstance is depicted in several descriptive data, such as the irregular distribution of SES, type of school, etc. Reason for this small sample was the negative stance of most school administration toward academic research in the school context.

Only few schools were willing to undergo the procedure of such a survey, and even less administrations were helpful concerning the execution of the survey as described in chapter 3.2.3. A huge majority of schools therefore refused to participate in this study, resulting in the relatively small sample as described. Reasons for this refusal may be found in the relatively high number of inquiries concerning surveys at school. Indeed, this was the most frequent yielded reason for a denial of cooperation. Moreover, schools may have refused to participate due to reasons of data security of their students. It was, however, never purpose of this study to use any data for any purposes other than the examination of the questions and hypotheses, which was made clear from the beginning of contacting school administrations. Other expressed reasons was the organizational and bureaucratic expense school administrations had at the beginning of the term.

Nevertheless, it seems that most schools – with some exceptions, which should be emphasized here – did not deem it worthy to participate in an academic study trying to investigate the subject of physical education classes.

As already pointed out throughout the whole study, attitude may be conceptualized in various ways. Moreover, for the present study, the original questionnaire developed by Subramaniam and Silvermann (2000) was further modified by the means of two subfactors, namely interaction and experience. For this reason, the means of all subfactors were calculated, as well (see Table 14). The attitudes toward teacher and curriculum were slightly lower than those toward interaction and experience. This suggests that students liked the togetherness and teamwork more than they liked the contents and the behavior of teachers in physical education classes. A very interesting implication of these findings is the highlighting of sports as being a field of enhanced and important social interaction, working as socialization facilitator. This agrees with basic claims of German sports education (cf. Balz & Kuhlmann, 2003, p. 118). The marginal difference between affective and cognitive conceptualization of attitude with the affective component having a better mean may indicate that, for 9th grade students, the affective component is more important than the cognitive one. Several authors have noted that children undergo dramatic changes and developments during adolescence, in which huge gaps between emotion and cognition can be observed, and in which the affective component starts to be regulated (cf. Steinberg, 2005; Choudhury, Blakemore, & Charman, 2006). The here presented data and sample may represent a stadium, in which this development has not been completed yet.

On the basis of the statistics, Hypothesis 1 was accepted, suggesting that students in general have positive attitudes toward physical education classes. This finding concurs with other German reports dealing with this subject (cf. Bräutigam, 2011b; Brettschneider & Kuhlmann, 2006; Digel, 1996; Kruber, 1996; Wydra, 2001). Still, the definition of a positive attitude may be a possible point of discussion and a reason for different outcomes or critique. The data analysis was therefore dependent on a certain point of reference. Thus, the question of what pertained as a positive attitude needed to be determined. In the present study, this value was set at relatively high level of 60 % of the total possible mean with 50 % constituting a neutral attitude. Hence, values only slightly differing from neutral attitudes were excluded from being positive. Another aspect concerning H1 was its formulation as one-sided. The p-value expounded in chapter 4.2, however, depicted the significance for a two-sided test. Yet, since this p-value was extremely low (0.000), the aspect of one-sided versus two-sided was neglected. The general positive attitudes of students toward physical education classes may result from their less formal nature compared to other subjects. Physical education lessons often provide the only possibilities for active movement during a school day (cf. Bräutigam, 2011b), which highlights its special status.

Hypothesis 2, on the contrary, was rejected due to the obtained data. Boys and girls do not seem to have different attitudes toward physical education classes. Even though these findings concur with the results of other international studies (cf. Omar-Fauzee et al., 2009; Stelzer et al., 2004), other studies report contrary findings (cf. Birtwistle & Brodie, 1991). These differences may result from different cultural environments or different grade levels. The present data, however, suggests that no differences exist for 14 to 16 year old students. It seems that the four subfactors curriculum, teacher, interaction, and experience satisfy the needs of both genders. This may be due to the fact that many schools teach physical education classes separated with respect to gender[51]. Thus, contents could be adjusted with regard to the target group, and therefore teachers are more likely to meet the needs of their students, who seem to have different interests (cf. Klewin, 1998). Moreover, the interaction and experience students gain during physical education classes may therefore not be subject to prudency based on the presence of the other gender, which may lead to more enjoyment and thus to similar attitudes.

---

[51] For a detailed discussion of coeducation in physical education classes see Wolters (2008).

The differences with respect to the school type in the present study were reported to be significant. To be precise, students of the Gymnasium and the Realschule had better attitudes than students of the Hauptschule. The distribution of the sample is of special interest, since only a minority of questioned students (20.8 %) visited the Hauptschule. This means that only a small sample of this type existed, which may not be representative. Nevertheless, H3 was accepted due to the data obtained. The differences between the Hauptschule and the other two types of school may as well be the result of the different demands of the school types.

However, no literature considering this question exists. It would therefore be of special interest to further examine students' attitudes with regard to the different types of schools. One possibility would be the additional investigation of hypotheses with the type of school as layer variable. Furthermore, the correlation of type of school and the SES may be of interest. Since the Hauptschule constitutes the type with the lowest demands, a higher rate of students with a bad SES may be possible.[52]

A further interesting issue in this context would be the general attitude of students toward school and toward exercise. A possible implication of the here presented results may be that students of the Hauptschule do in general have more negative attitudes toward school and toward exercise. Yet, as indicated above, these assumptions need to be further investigated. Concerning the SES as questioned in this survey, students with a good SES had better attitudes toward physical education classes than their mates with a bad SES. The dependence of students' attitudes on the SES – as suggested by the present study – may be due to the weak social and cultural capital students with a lower SES can draw on. Thus, school in general or particular subjects may be less interesting

---

[52] The present study suggests this possibility: 41.5 % of the students of the present sample visiting a Hauptschule was assigned a bad SES. For the Realschule and the Gymnasium, these were only 19.49 % and 11.02 %, respectively.

for those students. However, these outcomes contradict other findings of physical education class research (Birtwistle & Brodie, 1991; Valdez, 1997).

A major point of discussion in this context is the classification of the SES, which may be the reason for different results as described. In the present study, students' problems with reciting their parents' jobs were observed during the supervision of the completion of the questionnaire. This concurs with the registered feedback of the pretest, which was neglected. Students' inaccuracies giving their parents' jobs may therefore result in wrong classifications of their SES. Besides, the chosen dichotomous classification beards the problem of generalization due to only two possible forms. The differences between the present study and the above mentioned surveys may also have been caused by the different cultural contexts: not only differences in attitudes, but also differences in the definition of good or bad SES may have been the reason for the different outcomes.

Hence, future research dealing with this issue should choose a more distinct classification of the SES such as the ISCO-88 as described by several authors (cf. Ganzeboom & Treiman, 1996; Elias, 1997). The present study furthermore suggested differences in students' attitudes with regard to a migration background. Students without such a background were reported to have better attitudes toward physical education classes than their mates with a migration background. Reasons for these differences might be the varying importance of sport in different cultures. Exercise may be more or less prestigious in different cultures. Hence, the attitudes toward physical education classes may reflect general attitudes toward exercise. Again, however, these findings do not concur with the outcomes of other studies (cf. Brustad, 1996; Valdez, 1997). As already pointed out concerning the SES, these different outcomes may be due to different cultural contexts or different grade levels. It is therefore an important task of German physical education class research to investigate students' attitudes more thoroughly to provide comparable data. By the same token, no differences in attitudes between under-/overweight and normal weight children were found. Even though no other studies investigating this connection exist, the findings suggest that the BMI and its classification as suggested by the World Health Organization do only superficially determine classes of weight. Since the BMI is dependent on the height and the weight, an atypical weight relatively fast leads to a classification as under-/overweight.

However, the classification of the BMI does not necessarily depict a state of physical illness or sportiness, since muscle is heavier than fat (cf. Saladin, 2007). All this and the here presented data therefore suggest that the BMI as classified by the World Health Organization is not suitable for such investigations. If no other possibilities for a variable for students' constitution exist, it rather seems advisable to investigate dependent variables with regard to an unclassified BMI.

An influence of the grade point average on the attitude toward physical education classes was rejected as not being significant. Besides, only a very weak correlation between grade point average and attitude toward physical education classes was reported. This weak correlation may be due to the priorities students set. Students with better attitudes toward physical education classes might place more emphasis on sporting activity than on general school achievement, which in turn results in medium grade point averages. Physical education might, on the other hand, not be as important for those students who are good at school. This speculation as well as the findings of the present study disagree with the outcomes of another study conducted in Germany, suggesting that students who exercise more frequent place more emphasis on school success (cf. Brinkhoff, 1998). These outcomes are not confirmed by the present study. One reason for these differences is the relatively small sample of the present study. It is therefore the task of future research to clarify these possible connections.

Concerning the connection of students' physical education class grades and their attitudes toward the subject, a medium correlation between the two variables was found. This agrees with older findings (cf. Blum & Stüber, 1972), but also with more recent results of German physical education class research (cf. Opper, 1996). This connection may result from a more positive attitude toward sporting activity in general. Those students who exercise more frequent may show more positive attitudes toward sporting activity than those students who exercise less frequent. Besides, students who exercise more frequent in their free time are more likely to get good physical education class grades. This might suggest that sporty adolescents get better physical education class grades, and that hence their positive attitude toward physical education classes result from their sporting activity and not from their physical education grades. Thus it seems that several factors are intertwined and therefore have to be examined interconnectedly.

As hypothesis 9 has shown, attitudes of students do in fact differ regarding exercise and physical activity behavior. Students who reported to be more active had better attitudes than mates exercising on very irregular basis. However, significant differences only existed between students exercising on a very irregular basis and all other groups. This suggests that already a medium amount of sporting activity gives rise to better attitudes toward physical education. In the present study, this amount was also enough to get better physical education class grades.[53] Again this concurs with the findings of Opper (1996). Thus, there might in fact be various factors concerning sport coming into play, determining students' attitudes toward physical education classes. Future research could therefore concentrate on these factors, examining the connections between exercise behavior, physical education grade, and attitude toward exercise as well as attitude toward physical education classes.

Hypothesis 10 is of interest in this context, as well. Even though parents are usually said to be agents of socialization (cf. Lange, 2005, p. 28), their influence and role model function in the context of exercise and physical activity seems to be marginal. As the data has shown, no differences concerning students' attitudes with respect to the exercise and physical activity behavior of students' parents existed. Moreover, a post-

---

[53] Appendix 5 gives a table with the physical education grade distribution as a function of exercise frequency.

statistical contingency table examining the exercise frequency of students and those of their parents did not show any conspicuous forms.[54]

One reason for this data may be the age of the parents. Since students were between 14 and 16, most parents are assumed to be between 40 and 45, an age, in which regeneration from exercise takes longer (cf. Müller-Wohlfahrt, Ueblacker, & Hänsel, 2010, p. 119) and thus exercise may be performed less frequently. Besides, students' parents may be occupied with other obligations such as job and family, which influences their exercise behavior. Parents might have, on the contrary, exercised more frequently when they were younger, thus still having positive attitudes toward physical activity and exercise, which are passed to their children. And since the questionnaire at hand asked students about their parents' exercise and physical activity levels, the reported and the actual activity levels may just as well drift apart.

Future research in this direction thus not only needs to investigate parents' activity behavior, but rather their actual attitudes toward exercise and physical activity. Taking into consideration the exercise and physical activity behavior of students' peers, again the differences concerning the attitudes of students between the groups with peers exercising on very irregular basis and those exercising more frequently (at least 3 times a week) were observed. These differences are similar to those of the students themselves. Hence, the data indicates several possibilities: firstly, the exercise behavior of students' peers influences their attitude toward physical education classes. Secondly, students' reports about the exercise behavior of their peers was oriented toward their own exercise behavior, since they perceived their peers as being similarly active. And thirdly, students choose their friends according to their interests and hobbies. Thus, adolescents exercising more frequently keep company with even-aged who show a similar exercise pattern. Here, again, a post-statistic contingency table suggests that

---

[54] See Appendix 5 for the corresponding table.

students and their peers show similar exercise behaviors, thus indicating rather option two or three than option one. This would mean that the positive attitudes of students with peers exercising more frequently do rather stem from their own exercise behavior than from that of their peers. This opens another field for future research, which has not been explored yet. Even though differences were found regarding exercise and physical activity behavior, the status of the latter needs to be discussed.

Physical activity is, as chapter 2 has shown, a relatively broad term. Thus, problems for students may have occurred in making out differences between the two concepts, even though examples were given. These problems were also observed during the supervision of the procedure of filling out the questionnaire. Hence, the data concerning this variable may be inaccurate to the extend that students did not know what to classify as physical activity. It should therefore be of advantage for future research to concentrate on one variable alone – maybe comprising both exercise and physical activity.

In the context of exercise frequency, the reasons, institutions and preventions from doing sports are interesting. As chapter 4.1 has shown, most students and their peers are members in a sports club or in a fitness studio. Only few students reported themselves or their peers not being members of any institutions. The parents, on the other hand, were to a huge extend reported not to be members of any club. Just as described in discussing the exercise frequency of parents, this may not be due to a general negative attitude toward exercise.

As to the reasons for exercise, chapter 4.1 revealed that some major reasons were the same for all three groups. This is an interesting finding, since some reasons for sporting activity do not seem to be dependent on the age, but are rather ever-present: health benefits and fun. Other reasons, however, are indeed dependent on the age of the actors, such as weight controlling, prevention of aging, and recreation on the one hand, and increasing one's achievement potential and meeting friends on the other hand. These reasons could give an impulse to teachers, helping to understand the reasons of students to engage in sports.

Considering the reasons preventing from doing more sports, it seems that those are relatively trivial. Lack of time, lack of pleasure, or diseases and injuries were the major reasons for all groups. Here, it should be the goal of physical education classes to counteract in particular the second reason, that is the lack of pleasure. If students are shown how to incorporate sporting activity in their every day lives and how to enjoy such activities, more students may be guided toward a more active lifestyle, which is also one of the main tasks of physical education classes as detected by sports pedagogy (cf. Balz & Kuhlmann, 2003).

6   Conclusion

The study at hand investigated students' attitudes toward physical education classes. For this purpose, the influence of various independent variables on the attitudes was examined. Independent variables were mostly chosen on the basis of already existing literature. The analysis of the obtained data revealed several interesting facts that may be used by physical education teachers and administrations in two ways: firstly as information on the actual location of physical education classes, and secondly as information and advice for future acting in those classes.

The study revealed that students had a positive attitude toward physical education in general, and that these attitudes were not dependent on gender, the BMI, and students' grade point averages.

Students' attitudes were, however, dependent on the types of school they visited, their origin, and their socioeconomic status. Teachers and authorities therefore need to be aware of the various socioeconomic variables of their students when planning, realizing, and evaluating physical education lessons.

Besides the socioeconomic factors, sport connected factors influenced students' attitudes to a considerable extend. Their physical education grades, but also their exercise behavior and the exercise behavior of their peers were reported to have an impact on students' attitudes. Moreover, students' physical activity and exercise behavior and the related behaviors of their parents and peers were examined.

The results revealed that three quarters of students were members of any sports club. At the same time, however, this implies that one quarter of students are not members of any sports clubs, fitness clubs, or cultural association.

As most students indicated health benefits, fun, and the meeting of friends as major factors for their engaging in sporting activities, these findings provide considerable information about how to involve them best in physical education classes.

Physical education teachers should therefore be aware of these reasons and make use of this information to improve the benefits and outcomes of physical education classes for all students. This information may as well be used to guide students toward a more active lifestyle.

Most students reported a lack of pleasure as major factors preventing them form doing more sports. The usage of the provided information concerning students' reasons for doing sports may lead to students adopting more positive attitudes toward exercise and thus more active lifestyles due to more fun.

Future investigations should be directed toward several fields of research: the dependence of students' attitudes on their exercise behavior, which may as well affect their physical education class grades. Moreover, the differences between the different types of school need to be explored more thoroughly, and with it various implications for other independent variables.

Additionally, future research should bother about students' constitutions and their attitudes toward physical education classes. It should, however, consider a more complex construction of the constitution than the BMI. According to focus group, future research may therefore focus on diverse focus groups from kindergarten through high school, developing age specific instruments.

Interventions for changing students' attitudes towards physical education have not been prominent in empirical studies so far. Since there is evidence only on physical activity behavior change for outside school activities (Agbuga, Xiang, & McBride, 2013; Theodorakis & Goudas, 1997), interventions designed to change students' attitudes towards physical education are still in need to be developed and evaluated. However,

Digelidis et al. (2003) successfully targeted the improvement of students' attitudes toward physical exercise and aspects of physical education in a one-year intervention. Adding to the body of evidence, Ilker and Demirhan (2013) reportet a positive effect of a mastery-oriented motivational climate on students' attitudes towards physical education.

Moreover, there is a lack of evidence regarding changes of attitudes towards physical education throughout adulthood. Although some studies assessed data on particular school career stages and grades (Dismore & Bailey, 2011; Ilker et al., 2011; Subramaniam & Silverman, 2007), students' attitude changes and its associated factors are unknown. Omar-Fauzee et al. (2009) presented data on attitudes towards physical education from a retrospective perspective in college students, but no data on attitude change and progress since their school days has been collected. Hence, longitudinal studies are needed that may also prove that positive attitudes towards physical education really transform into a physically active lifestyle afterwards and reveal confounders in the process. Suitable theoretic frameworks for attitude change interventions and investigations may be found in the *Learning Theory of Attitude Change, Elaboration Likelihood Theory of Attitude Change*, or the *Dissonance Theory of Attitude Change*, as these theories have been prominent in social psychology research regarding attitude change (McGuire, 1985; Smith & Mackie, 2007).

This study thus revealed several implications and information about the attitudes of students toward physical education classes and their sporting activity. It is the task of future research to further explore this field of research and to provide additional information for teachers and authorities to help making physical education classes more attractive and effective.

References

Aaron, D.J., Storti, K.L., Robertson, R.J., Kriska, A.M., & LaPorte, R.E. (2002). Longitudinal Study of the Number and Choice of Leisure Time Physical Activities from Mid to Late Adolescence. Implications for School Curricula and Community Recreation Programs. *Archives of Pediatrics and Adolescent Medicine, 156*, 1057-1080.

Agbuga, B., Xiang, P., & McBride, R. (2013). Students' attitudes toward an after-school physical activity programme. *European Physical Education Review, 19*(1), 91-109.

Al-Liheibi, A.H.N. (2008). *Middle and High School Students' Attitudes Toward Physical Education in Saudi Arabia.* Doctoral Dissertation, University of Arkansas.

Allport, G. (1935). Attitudes. In C. Murchison (Ed.), *A Handbook of Social Psychology* (pp. 789-844). Worcester, MA: Clark University Press.

Ajzen, I. (1985). From Intentions to Actions: A Theory of Planned Behavior. In J. Kuhl & J. Beckmann (Eds.), *Action Control: From Cognitions to Behavior* (pp. 11-39). New York, NY: Springer.

Ajzen, I. (1989). Attitude Structure and Behavior. In A. Pratkanis, S. Breckler & A. Greenwald (Eds.), *Attitude Structure and Function* (241-274). New Jersey: Lawrence Erlbaum Associates.

Ajzen, I. (2007). *Attitudes, Personality, and Behavior. Mapping social psychology.* New York: McGraw-Hill International.

Anderssen, N. (1993). Perception of physical education classes among young adolescents: do physical education classes provide equal opportunities to all students. *Health Education Research, 8*(2), 167-179.

Armstrong, N., & Bray, S. (1991). Physical activity patterns defined by continuous heart rate monitoring. *Archives of Disease in Childhood, 66*, 245-247.

Arabaci, R. (2009). Attitudes toward physical education and class preferences of Turkish secondary and high school students. *Elementary Education Online, 8*(1), 2-8.

Bagozzi, R.P., & Burnkrant, R.E. (1979). Attitude measurement and behavior change: A reconsideration of attitude organization and its relationship to behavior. *Advances in Consumer Research, 6*(1), 295-302.

Bailey, R. (2006). Physical Education and Sport in Schools: A Review of Benefits and Outcomes. *Journal of School Health, 76*(8), 397-401.

Balnaves, M., & Caputi, P. (2001). *Introduction to Quantitative Research Methods. An Investigative Approach.* London: SAGE.

Balz, E., Bräutigam, M., Miethling, W., & Wolters, P. (Eds.) (2011). *Empirie des Schulsports.* Aachen: Meyer & Meyer.

Balz, E., & Kuhlmann, D. (2003). *Sportpädagogik: Ein Lehrbuch in 14 Lektionen.* Aachen: Meyer & Meyer.

Balz, E., & Neumann, P. (1999). Erziehender Sportunterricht. In W. Günzel & R. Laging (Eds.), *Neues Taschenbuch des Sportunterrichts* (p. 162-192). Baltmannsweiler: Schneider.

Bernstein, E., Phillips, S.R., & Silverman, S. (2011). Attitudes and Perceptions of Middle School Students Toward Competitive Activities in Physical Education. *Journal of Teaching in Physical Education, 30*(1), 69-83.

Bibik, J.M., Goodwin, S.C., & Orsega-Smith, E.M. (2007). High School Students' Attitudes Toward Physical Education in Delaware. *The Physical Educator, 64*(4), 192-204.

Biddle, S., Sallis, J.F., & Cavill, N. (Eds.) (1998). *Young and Active? Young People and Health-Enhancing Physical Activity – Evidence and Implications.* London: Health Education Authority.

Biddle, S.J.H., & Wang, C.K. (2003). Motivation and self-perception profiles and links with physical activity in adolescent girls. *Journal of Adolescence, 26*(6), 687-701.

Birtwistle, G.E., & Brodie, D.A. (1991). Children's attitudes towards activity and perceptions of physical education. *Health Education Research, 6*(4), 465-478.

Blum, P., & Stüber, J. (1972). Der Einfluss der Sportnote auf die Motivation der Schüler im Sportunterricht. *Die Leibeserziehung, 21*(11), 376-382.

Bohner, G. (2002). Einstellungen. In W. Stroebe, K. Jonas & M. Hewstone (Eds.), *Sozialpsychologie: Eine Einführung* (4th ed,) (pp. 265-318). Berlin: Springer.

Bohner, G., & Wänke, M. (2002). *Attitudes and Attitude Change.* Hove, East Sussex: Psychology Press.

Bräutigam, M. (2011a). *Sportdidaktik: Ein Lehrbuch in 12 Lektionen* (4th ed.). Aachen: Meyer & Meyer.

Bräutigam, M. (2011b). Schülerforschung. In E. Balz, M. Bräutigam, W. Miethling & P. Wolters (Eds.), *Empirie des Schulsports* (pp. 65-94). Aachen: Meyer & Meyer.

Brettschneider, W.-D., & Kuhlmann, D. (2006). Die Schulsportuntersuchung und ihre modulare Struktur – von der Entstehung bis zur SPRINT-Studie. In Deutscher Sportbund (Ed.), *Die SPRINT-Studie. Eine Untersuchung zur Situation des Schulsports in Deutschland* (pp. 3-11). Aachen: Meyer & Meyer.

Brinkhoff, K.-P. (1998). *Sport und Sozialisation im Jugendalter. Entwicklung, soziale Unterstützung und Gesundheit*. Weinheim: Juventa.

Brustad, R.J. (1996). Attraction to Physical Activity in Urban Schoolchildren: Parental Socialization and Gender Issues. *Research Quarterly for Exercise and Sport, 67*(3), 316-323.

Carlson, T.B. (1995). We hate gym: Student alienation from physical education. *Journal of Teaching in Physical Education, 14*(4), 467-477.

Cavill, N., Biddle, S., & Sallis, J.F. (2001). Health enhancing physical activity for young people: Statements of the United Kingdom Expert Consensus Conference. *Pediatric Exercise Science, 13*, 12-15.

Caspersen, C., Powell, K., & Christenson, G. (1985). Physical Activity, Exercise, and Physical Fitness: Definitions and Distinctions for Health-Related Research. *Public Health Reports, 100*, 126-131.

Chatterjee, S. (2013). Attitudes toward Physical Education of School Going Adolescents in West Bengal. *International Journal of Innovative Research in Science, Engineering and Technology, 2*(11), 6068-6073.

Chatterjee, S., Nandy, S., & Adhikari, S. (2012). Impact of Sports Perfectionism on Development of Attitude towards Physical Education of the School-Going Adolescents. *IOSR Journal of Humanities and Social Science, 2*(3), 46-51.

Cheng, L.A., Mendonca, G., & de Farias Junior, J.C. (2014). Physical activity in adolescents: analysis of the social influence of parents and friends. *Journal de Pediatria, 90*(1), 35-41.

Cohen, L., Manion, L., & Morrison, K. (2011). *Research Methods in Education* (7th ed.). London; New York: Routledge.

Colquitt, G., Walker, A., Langdon, J.L., McCollum, S., & Pomazal, M. (2012). Exploring Student Attitudes Toward Physical Educatio and Implicatios for Policy. *Sport Scientific Practical Aspects, 9*(2), 5-12.

Coulter, M., & Woods, C.B. (2011). An exploration of children's perceptions and enjoyment of school-based physical activity and physical education. *Journal of Physical Activity & Health, 8*(5), 645-654.

Crespo, N.C., Corder, K., Marshall, S., Norman, G.J., Patrick, K., Sallis, J.F., & Elder, J.P. (2013). An examination of multilevel factors that may explain gender differences in children's physical activity. *Journal of Physical Activity & Health, 10*(7), 982-992.

Crisp, R.J., & Turner, R.N. (2007). *Essential Social Psychology*. London: Sage Publications.

Choudhury, S., Blakemore, S.-J., & Charman, T. (2006). Social Cognitive Development during Adolescence. *Social Cognitive and Affective Neuroscience, 1*, 165-174.

Deci, E.L., & Ryan, R.M. (1985). *Intrinsic motivation and self-determination in human behavior.* New York, NY: Plenum.

Deutsche Sportjugend (DSJ) (2009). *Stellungnahme des dsj-Vorstands zum Zweiten Deutschen Kinder- und Jugendsportberichts.* Retrieved September 26, 2013 from: http://www.lsvbw.de/cms/docs/doc6863.pdf

Deutscher Sportbund (DSB) (Ed.). (2006). *Die SPRINT-Studie. Eine Untersuchung zur Situation des Schulsports in Deutschland.* Aachen: Meyer & Meyer.

Digel, H. (1996). Schulsport – wie ihn Schüler sehen. Eine Studie zum Schulsport in Südhessen (Teil 1). *Sportunterricht, 45*(8), 324-339.

Digelidis, N., Papaioannou, A., Laparidis, K., & Christodoulidis, T. (2003). A one-year intervention in 7th grade physical education classes aiming to change motivational climate and attitudes toward exercise. *Psychology of Sport and Exercise, 4*(3), 195-210.

Dismore, H., & Bailey, R. (2011). Fun and enjoyment in physical education: young people's attitudes. *Research Papers in Education, 26*(4), 499-516.

Döhring, V. & Gissel, N. (2009). Planung und Auswertung von Sportunterricht. In H. Lange & S. Sinning (Eds.), *Handbuch Sportdidaktik* (2nd ed.) (pp. 426-446). Balingen: Splitta.

Dyson, B. (2006). Students' perspectives of physical education. In D. Kirk, D. Macdonald & M. O'Sullivan (Eds.), *The handbook of physical education* (pp. 327-346). London: Sage.

Eagly, A.H., & Chaiken, S. (1998). Attitude Structure and Function. In D.T. Gilbert, S.T. Fisk & G. Lindsey (Eds.), *Handbook of Social Psychology* (pp. 269-322). New York, NY: McGraw-Hill.

Elias, P. (1997). Occupational Classification (ISCO-88): Concepts, Methods, Reliability, Validity and Cross-National Comparability. *OECD Labour Market and Social Policy Occasional Papers*, 20, OECD Publishing. Retrieved September 26, 2013 from: http://www.oecd-ilibrary.org/docserver/download/5lgsjhvj7td8.pdf?expires=1355260972&id=id&accname=guest&checksum=F13A84573

Ennis, C.D. (1996). Students' experiences in sport-based physical education: [More than] apologies are necessary. *Quest, 48*(4), 453-456.

Fairclough, S., & Stratton, G. (2005). 'Physical education makes you fit and healthy'. Physical education's contribution to young people's physical activity levels. *Health Education Research, 20*(1), 14-23.

Fazio, R.H., & Towles-Schwen, T. (1999). The MODE Model of Attitude-Behavior Processes. In S. Chaiken & Y. Trope (Eds.), *Dual Process Theories in Social Psychology.* New York, NY: Guilford Press.

Fazio, R.H., & Zanna, M.P. (1981). Direct experience and attitude-behavior consistency. In L. Berkowitz (Ed.), *Advances in experimental social psychology* (pp. 161-202). New York, NY: Academic.

Fishbein, M., & Ajzen, I. (1975). *Belief, attitude, intention, and behavior: an introduction to theory and research*. Reading, MA: Addison-Wesley.

Flanagan, J.C. (1954). The critical incident technique. *Psychological Bulletin, 51*(4), 327-358.

Fox, K.R. (1998). Advances in the Measurement of the Physical Self. In J.L. Duda (Ed.), *Advances in Sport and Exercise Psychology Measurement* (pp. 295-310). Morgantown, WV: Fitness Information Technology.

Gardner, R.C. (1985). *Social Psychology and Second Language Learning. The Role of Attitudes and Motivation*. London: Arnold.

Gage, N., & Berliner, D. (1996). *Pädagogische Psychologie* (5th ed.). Weinheim: Beltz.

Ganzeboom, H.B.G., & Treiman D.J. (1996). Internationally Comparable Measures of Occupational Status for the 1988 International Standard Classification of Occupations. *Social Science Research, 25*, 201-239.

Gardner, R. (1985). *Social Psychology and Second Language Learning. The Role of Attitudes and Motivation*. London: Arnold.

Gebken, U. (2005). Guter Sportunterricht. Merkmale und Ratschläge für die Praxis. *Praxis in Bewegung, Sport und Spiel, 5*(1), 38-41.

Gerlach, E., Kussin, U., Brandl-Bredenbeck, H.-P., & Brettschneider, W.-D. (2006). Der Sportunterricht aus Schülerperspektive. In Deutscher Sportbund (Ed.), *Die SPRINT-Studie. Eine Untersuchung zur Situation des Schulsports in Deutschland* (pp. 107-144). Aachen: Meyer & Meyer.

Gieß-Stüber, P., Neuber, N., Gramespacher, E., & Salomon, S. (2009). Mädchen und Jungen im Sport. In W. Schmidt (Ed.), *Zweiter Deutscher Kinder- und Jugendsportbericht. Schwerpunkt: Kindheit* (2nd ed.) (pp. 63-83). Schorndorf: Hofmann.

Goktas, Z. (2012). The Attitudes of Physical Education and Sport Students towards Information and Communication Technologies. *TechTrends: Linking Research and Practice to Improve Learning 56*(2), 22-30.

Graber, K.C. (2001). Research on teaching in physical education. In V. Richardson (Ed.), *Handbook on research on teaching* (pp. 491-519). Washington, DC: American Educational Research Association.

Green, K., & Lamb, K. (2000). Health-Related Exercise, Effort Perception and Physical Education. *European Journal of Physical Education, 5*(1), 88-103.

Hagendorf, H., Müller H., Krummenacher, J., & Schubert, T. (2011). *Wahrnehmung und Aufmerksamkeit. Allgemeine Psychologie für Bachelor*. Berlin: Springer.

Haynes, J., Fletcher, T., & Miller, J. (2008). Does Grouping By Perceived Ability Sustain Student Attitude Towards Physical Education? In P.L. Jeffery (Ed.), *2008 Australian Association for Research in Education (AARE) Conference Papers Collection* [CD-ROM]. Coldstream, VIC: AARE.

Hagger, M., Cale, L., Almond, L., & Krüger, A. (1997). Children's Physical Activity Levels and Attitudes toward Physical Activity. *European Physical Education Review, 3*(2), 144-164.

Hardman, K. (1998). The Fall and Rise of School Physical Education in International Context. In R. Naul, K. Hardman, M. Piéron & B. Skirstad (Eds.), *Physical Activity and Active Lifestyle of Children and Youth* (pp. 89-107). Schorndorf: Hofmann.

Harris, J., & Cale, L. (1997). Activity Promotion in Physical Education. *European Physical Education Review, 3*(1), 58-67.

Hartmann-Tews, I., & Luetkens, S.A. (2003). Jugendliche Sportpartizipation und somatische Kulturen aus Geschlechterperspektive. In W. Schmidt, I. Hartmann-Tews & W.-D. Brettschneider (Eds.), *Erster Deutscher Kinder- und Jugendsportbericht* (pp. 297-318). Schorndorf: Hofmann.

Hausmann, C. (2009). *Psychologie und Kommunikation für Pflegeberufe. Ein Handbuch für Ausbildung und Praxis* (2nd ed.). Wien: Facultas.

Heemsoth, T., & Miethling, W. (2012). Schülerwahrnehmungen des Unterrichtsklimas. Entwicklung eines Fragebogens und Befunde zum Sportunterricht. *Sportwissenschaft, 42*(4), 228-239.

Hockenbury, D., & Hockenbury, S.E. (2007). *Discovering Psychology*. New York, NY: Worth Publishers.

Hoffmann, B., Martini, H., Martini, U., Rebel, G., Wickel, H., & Wilhelm, E. (Eds.) (2004). *Gestaltungspädagogik in der sozialen Arbeit*. Paderborn: Schöningh.

Hoffmeyer-Zlotnik, J.H.P., & Geis, A.J. (2003). Berufsklassifikation und Messung des beruflichen Status/Prestige. *ZUMA-Nachrichten, 27*, 125-138.

Hogg, M., & Vaughan, G. (2008). *Social Psychology* (5th ed.). Harlow, Essex: Pearson Education.

Howie, E.K., & Pate, R.R. (2012). Physical activity and academic achievement in children: A historical perspective. *Journal of Sport and Health Science, 1*(3), 160-169.

Ilker, G., Arslan, Y., & Demirhan, G. (2011). An Examination of Turkish High School Students' Attitudes toward Physical Education with Regard to Gender and Grade Level. *Sport si Societate, 11*(2), 91-95.

Ilker, G., & Demirhan, G. (2012). The effects of different motivational climates on students' achievement goals, motivational strategies and attitudes toward physical education. *Educational Psychology, 33*(1), 59-74.

Jackson, A., Morrow J., Hill, D. & Dishman, R. (Eds.) (2004). *Physical Activity for Health and Fitness* (updated edition). Champaign, IL: Human Kinetics.

Johns, G., & Saks, A.M. (2008). *Organizational behaviour. Understanding and managing life at work* (7th ed.). Toronto: Prentice Hall.

Kastrup, V. (2009). *Der Sportlehrerberuf als Profession. Eine empirische Studie zur Bedeutung des Sportlehrerberufs.* Schorndorf: Hofmann.

Keating, X.D., Silverman, S., & Kulinna, P.H. (2003). Preservice Physical Education Teacher Attitudes toward Fitness Tests and the Factors Influencing their Attitudes. *Journal of Teaching in Physical Education, 21*(2), 193-207.

Kenyon, G.S. (1968). Six scales for assessing attitude toward physical activity. *Research Quarterly for Exercise and Sport, 39*(3), 566-574.

Kirkcaldy, B.D., & Shephard, R.J. (1990). Therapeutic aspects of leisure and sport. *International Journal of Sport Psychology, 21*(3), 165-184.

Kirkcaldy, B.D., Shephard, R.J., & Siefen, R.G. (2002). The relationship between physical activity and self-image and problem behavior among adolescents. *Social Psychiatry and Psychiatric Epidemiology, 37*(11), 544-550.

Klewin, G. (1998). Mädchen und Jungen im Schulsport. In K. Behm & K. Petzsche (Eds.), *Mädchen und Jungen im Schulsport. Landesweite Fachtagung im Rahmen des Landesprogramms der Landesregierung NRW „Mehr Chancen für Mädchen und Frauen im Sport"* (pp. 88-95). Bönen: Kettler.

Koch-Priewe, B., Niederbacher, A., Textor, A., & Zimmermann, P. (2009). *Jungen – Sorgenkinder oder Sieger? Ergebnisse einer quantitativen Studie und ihre pädagogischen Implikationen.* Wiesbaden: VS Verlag.

Kolb, B., & Whishaw, I.Q. (2009). *Fundamentals of human neuropsychology* (6th ed.). New York, NY: Worth.

Kretschmann, R. (2015, im Druck). Pupils' and studets's attitudes towards physical education: A review. *International Journal of Physical Education, 16*(2), xx-xx.

Kretschmann, R., & Wrobel, D. (2014). Students' Attitudes towards Physical Education. In R. Todaro (Ed.), *Handbook of Physical Education Research: Role of School Programs, Children's Attitudes and Health Implications* (pp. 1-24). New York, NY: Nova Science Publishers.

Krieger, C. (2006). Zur Rekonstruktion sportunterrichtlicher Situationen aus Schüler- und Lehrersicht. In M. Kolb (Ed.), *Empirische Schulsportforschung* (p. 60-72). Hohengehren: Schneider.

Kromrey, H. (2006). *Empirische Sozialforschung* (11th ed.). Stuttgart: Lucius & Lucius.

Kruber, D. (1996). Lieblingsfach Sport. *Sportunterricht, 45*(1), 4-8.

Kurz, D. (1990). *Elemente des Schulsports* (3rd ed.). Schorndorf: Hofmann.

Kurz, D. (2004). Von der Vielfalt sportlichen Sinns zu den pädagogischen Perspektiven im Schulsport. In P. Neumann & E. Balz (Eds.), *Mehrperspektivischer Sportunterricht* (pp. 57-70). Schorndorf: Hofmann.

Lamnek, S. (2005). *Qualitative Sozialforschung: Lehrbuch* (4th ed.). Weinheim: Beltz.

Lampert, T., Mensink, G.B., Romahn, N., & Woll, A. (2007). Körperlich-Sportliche Aktivität von Kindern und Jugendlichen in Deutschland. Ergebnisse des Kinder- und Jugendgesundheitssurveys (KiGGS). *Bundesgesundheitsblatt – Gesundheitsforschung – Gesundheitsschutz, 5/6*, 634-642.

Lange, E. (2005). *Soziologie des Erziehungswesens. Studienskript zur Soziologie* (2nd ed.). Wiesbaden: VS Verlag.

Leiße, O., Buhl, T., Leiße, U., & Berger U. (Eds.) (2006). *Psychologie und Soziologie: Lehr- und Lernbuch für die Verwaltung.* München: Oldenbourg.

Locke, H.S., & Braver, T.S. (2008). Motivational influences on cognitive control: behavior, brain activation, and individual differences. *Cognitive, Affective, & Behavioral Neuroscience, 8*(1), 99-112.

Luke, M.D., & Sinclair, G.D. (1991). Gender Differences in Adolescents' Attitudes Toward School Physical Education. *Journal of Teaching in Physical Education, 11*(1), 31-46.

Macdonald, D., Rodger, S., Ziviani, J., Jenkins, D., Batch, J., & Jones, J. (2004). Physical activity as a dimension of family life for lower primary school children. *Sport, Education and Society, 9*(3), 307-325.

Madden, T.J., Ellen, P.S., & Ajzen, I. (1992). A Comparison of the Theory of Planned Behavior and the Theory of Reasoned Action. *Personality and Social Psychology Bulletin, 18*(1), 3-9.

Maderthaner, R. (2008). *Psychologie.* Stuttgart: UTB.

Maiano, C., Ninot, G., & Bilard, J. (2004). Age and gender effects on global self-esteem and physical self-perception in adolescents. *European Physical Education Review, 10*(1), 53-69.

Marsh, H.W. (1998). Age and Gender Effects in Physical Self-Concepts for Adolescent Elite Athletes and Nonathletes. A Multicohort-Multioccasion Design. *Journal of Sport and Exercise Psychology, 20*(3), 237-259.

Mason, V. (1995). *Young People and Sport.* London: Sports Council.

McGuire, W.J. (1985). Attitudes and attitude change. In G. Lindzey & E. Aronson (Eds.), *Handbook of Social Psychology* (3rd ed.) (pp. 233-346). Ney York, NY: Random House.

McGuire, W.J. (1989). The structure of individual attitudes and attitude systems. In A. Pratkanis, S. Breckler & A. Greenwald (Eds.), *Attitude structure and function* (pp. 37-69). Hillsdale, NJ: Erlbaum.

Meinefeld, W. (1977). *Einstellung und soziales Handeln*. Reinbek: Rowohlt.

Mietzel, G. (2002). *Wege in die Entwicklungspsychologie. Kindheit und Jugend.* Weinheim: Belz.

Ministerium für Kultus, Jugend und Sport Baden-Württemberg (2005). *Ergänzende Informationen zur DSB-SPRINT Studie.* Retrieved September 26, 2013 from: http://www.kultusportal-bw.de/servlet/PB/show/1209519/ LIS_sprint.pdf

Mohammed, H.R., & Mohammad, M.A. (2012). Students Opinions and Attitudes towards Physical Education Classes in Kuwait Public Schools. *College Student Journal, 46*(3), 550-566.

Müller-Wohlfahrt, H.W., Ueblacker, P., & Hänsel L. (2010). *Muskelverletzungen im Sport.* Stuttgart: Thieme.

Myers, D.G. (2012). *Social Psychology* (11th ed.). New York, NY: McGraw-Hill.

National Association for Sport and Physical Education (NASPE). (2004). *Moving into the Future. National standards for physical education* (2nd ed.). Reston: McGraw-Hill.

Neumann, P., & Balz, E. (Eds.) (2004). *Mehrperspektivischer Sportunterricht. Orientierungen und Beispiele.* Schorndorf: Hofmann.

Ntoumanis, N. (2001). A self-determination approach to the understanding of motivation in physical education. *British Journal of Educational Psychology, 71*(2), 225-242.

Omar-Fauzee, M.S., Jamalis, M., Yusof, A., Zarina, M., Omar, R., Padli, H., Norazemi, A., Junaidi, A., Dewi, A.M., Latif, R.A., Johar, M., & Nasaruddin, M.N. (2009). College Students Perception on Physical Education Classes During their High School Days. *European Journal of Social Sciences, 7*(4), 69-76.

Oppenheim, A. (2000). *Questionnaire Design, Interviewing and Attitude Measurement* (2nd ed.). London: Continuum International Publishing Group.

Opper, E. (1996). Erleben Mädchen den Schulsport anders als Jungen? *Sportunterricht, 45*(8), 349-356.

Orunaboka, T.T. (2011). Attitude of Nigeria Secondary School Students towards Physical Education as a Predictor of Achievement in the Subject. *Journal of Education and Practice, 2*(6), 71-77.

Page, A., Ashford, B., Fox, K., & Biddle, S. (1993). Evidence of cross-cultural validity for the physical self-perception profile. *Personality and Individual Differences, 14*(4), 585-590.

Pano, G., & Markola, L. (2011). 14-18 years old children attitudes, perception and motivation towards extra curricular physical activity and sports. *Journal of Human Sport and Exercise, 7*(1Proc), 51-66.

Park, S.Y. (1995). *Identifying the Attitudes of Students in Large Urban High Schools to Physical Education*. Unpublished Master's Thesis, California State University, Los Angeles.

Pate, R., Heath, G., Dowda, M., & Trost, S. (1996). Associations between Physical Activity and Other Health Behaviors in a Representative Sample of US Adolescents. *American Journal of Public Health, 86*(11), 1577-1581.

Pelletier, L.G., Tuson, K.M., Fortier, M.S., Vallerand, R.J., Briere, N.M., & Blais, M.R. (1995). Toward a new measure of intrinsic motivation, extrinsic motivation, and amotivation in sports: The Sports Motivation Scale (SMS). *Journal of Sport and Exercise Psychology, 17*(1), 35-53.

Phillips, S.R., & Silverman, S. (2012). Development of an Instrument to Assess Fourth and Fifth Grade Students' Attitudes toward Physical Education. *Measurement in Physical Education and Exercise Science, 16*(4), 316-327.

Pickens, J. (2005). Attitudes and Perceptions. In N. Borkowsky (Ed.), *Organizational Behavior in Health Care* (pp. 43-76). Sudbury, MA: Jones and Bartlett Publishers.

Pomerantz, J.R. (2003). Perception: Overview. In L. Nadel (Ed.), *Encyclopedia of Cognitive Science* (Vol. 3) (pp. 527-537). London: Nature Publishing Group.

Portman, P.A. (1995). Who is having fun in physical education classes? Experiences of six-grade students in elementary and middle schools. *Journal of Teaching in Physical Education, 14*(4), 445-453.

Prochaska, J.J., Sallis, J.F., & Long, B. (2001). A physical activity screening measure for use with adolescents in primary care. *Archive of Pediatric Adolescent Medicine, 155*, 554-559.

Prochaska, J.J., Sallis, J.F., Slymen, D.J., & McKenzie, T.L. (2003). A longitudinal study of children's enjoyment of physical education. *Pediatric Exercise Science, 15*, 170-178.

Raustorp, A., Stahle, A., Gudasic, H., Kinnunen, A., & Mattsson, E. (2005). Physical activity and self-perception in school children assessed with the Children and Youth - Physical Self-Perception Profile. *Scandinavian Journal of Medicine and Science in Sports, 15*(2), 126-134.

Reilly, J.J., Penpraze, V., Hislop, J., Davies G., Grant,S., & Paton, J.Y. (2008). Objective measurement of physical activity and sedentary behavior: review with new data. *Archives of Disease in Childhood, 93*, 614-619.

Rethorst, S., Fleig, P., & Willimczik, K. (2009). Effekte motorischer Förderung im Kindergarten. In W. Schmidt (Ed.), *Zweiter Deutscher Kinder- und Jugendsportbericht. Schwerpunkt: Kindheit* (2nd ed.) (pp. 237-254). Schorndorf: Hofmann.

Rikard, G.L., & Banville, D. (2006). High school student attitudes about physical education. *Sport, Education and Society, 11*(4), 385-400.

Roberts, K. (1995). School children and sport. In L. Lawrence, E. Murdoch & S. Parker (Eds.), *Professional and Development Issues in Leisure, Sport and Education* (pp. 337-348). Brighton: LSA Publications.

Robinson, H. (1994). *Perception. Problems of Philosophy*. London: Routledge.

Ryan, S., Fleming, D., & Maina, M. (2003). Attitudes of Middle School Students Toward their Physical Education Teachers and Classes. *The Physical Educator, 60(*2), 28-42

Saladin, K. (2007). *Anatomy and Physiology. The Unity of Form and Function* (4. edition). New York: McGraw-Hill International.

Salvy, S.-J., Wojslawowicz Bowker, J., Roemmich, J.N., Romero, N., Kieffer, E., Paluch, R., & Epstein, L.H. (2007). Peer Influence on Children's Physical Activity: An Experience Sampling Study. *Journal of Pediatric Psycholology, 33*(2), 39-49.

Sanes, L.A.C. (2009). Students' Attitudes Towards Physical Education. *La Sallian Research Forum, 14*(4).

Scharfetter, C. (2010). *Allgemeine Psychopathologie: Eine Einführung* (6th ed.). Stuttgart: Thieme.

Scherler, K. (2000). Sport als Schulfach. In P. Wolters (Ed.), *Didaktik des Schulsports* (pp. 36-60). Schorndorf: Hofmann.

Schmidt, W., Zimmer R., & Völker, K. (2009). Zweiter Deutscher Kinder- und Jugendsportbericht. Schwerpunkt: Kindheit. Zusammenfassung, Forschungsdesiderate und Handlungsempfehlungen. In W. Schmidt (Ed.), *Zweiter Deutscher Kinder- und Jugendsportbericht. Schwerpunkt: Kindheit* (2nd ed.) (pp. 467-476). Schorndorf: Hofmann.

Schneider, W., & Büttner, G. (2002). Entwicklung des Gedächtnisses bei Kindern und Jugendlichen. In R. Oerter & L. Montada (Eds.), *Entwicklungspsychologie* (pp. 495-516). Weinheim: Belz.

Seel, N. (2003). *Psychologie des Lernens: Lehrbuch für Pädagogen und Psychologen* (2nd ed.). Stuttgart: UTB.

Serwe, E., & Thiele, J. (2008). Aktuelle Themenfelder der Schulsportentwicklung. In Dortmunder Zentrum für Schulsportforschung (Ed.), *Schulsportforschung: Grundlagen, Perspektiven und Anregungen* (pp. 154-170). Aachen: Meyer & Meyer.

Sheppard, B.H., Hartwick, J., & Warshaw, P.R. (1988). The Theory of Reasoned Action: A Meta-Analysis of Past Research with Recommendations for Modifications and Future Research. *Journal of Consumer Research,, 15*(3), 325-343.

Silverman, S. (1996). A pedagogical model of human performance determinants in sports *Proceedings of the pre-congress symposium of the 1996 Seoul International Sport Science Congress* (Vol. I) (pp. 476-488). Seoul, Korea: Korean Alliance for Health, Physical Education, Recreation and Dance.

Silverman, S., & Subramaniam, P.R. (1999). Student Attitude Toward Physical Education and Physical Activity: A Review of Measurement Issues and Outcomes. *Journal of Teaching in Physical Education, 19*(1), 97-125.

Simon, J.A., & Smoll, F.L. (1974). An instrument for assessing children's attitudes toward physical activity. *Research Quarterly, 45*(4), 407-415.

Simons-Morton, B.G. (1994). Implementing health-related physical education. In R.R. Pate & R.C. Hohn (Eds.), *Health and Fitness through Physical Education* (pp. 137-146). Champaign, IL: Human Kinetics.

Sleap, M., & Warburton, P. (1992). Physical activity levels of 5-11 year old children in England as determined by continuous observation. *Research Quarterly for Exercise and Sport, 63*(3), 238-245.

Smith, E.R., & Mackie, D.M. (2007). *Social Psychology*. London: Psychology Press.

Solmon, M.A. (2006). Learner cognition. In D. Kirk, D. Macdonald & M. O'Sullivan (Eds.), *The handbook of physical education* (pp. 226-241). London: Sage.

Solmon, M.A., & Carter, J.A. (1995). Kindergarten and first-grade students' perceptions of physical education in one teacher's classes. *Elementary School Journal, 95*(4), 355-365.

Standage, M., Duda, J.L., & Ntoumanis, N. (2003). A Model of Contextual Motivation in Physical Education: Using Constructs from Self-Determination and Achievement Goal Theories to Predict Physical Activity Intentions. *Journal of Educational Psychology, 95*(1), 97-110.

Steinberg, L. (2005). Cognitive and Affective Development in Adolescence. *TRENDS in Cognitive Science, 9*(2), 69-74.

Stelzer, J., Ernest, J.M., Fenster, M.J., & Langford, G. (2004). Attitudes toward Physical Education. A Study of High School Students from Four Countries - Austria, Czech Republic, England, and USA. *College Student Journal, 38*(2), 171-179.

Sternberg, R.J., & Sternberg, K. (2012). *Cognitive psychology* (6th ed.). Belmont, CA: Wadsworth.

Stibbe, G. (2011). Lehrplanarbeit im Fach Sport. In E. Balz, M. Bräutigam, W. Miethling & P. Wolters (Eds.), *Empirie des Schulsports* (pp. 197-207). Aachen: Meyer & Meyer.

Subramaniam, P.R., & Silverman, S. (2000). Validation of Scores from an Instrument Assessing Student Attitude Toward Physical Education. *Measurement in Physical Education and Exercise Science, 4*(1), 29-43.

Subramaniam, P.R., & Silverman, S. (2002). Using complimentary data: An investigation of student attitude in physical education. *Journal of Sport Pedagogy, 8*, 74-91.

Subramaniam, P.R., & Silverman, S. (2007). Middle School Students' Attitudes Toward Physical Education. *Teaching and Teacher Education, 23*(5), 602-611.

Suminski, R.R., Petosa, R., Utter, A.C., & Zhang, J.J. (2002). Physical Activity among Ethnically Diverse College Students. *Journal of American College Health, 51*(2), 75-80.

Tannehill, D., Romar, J.-E., O'Sullivan, M., England, K., & Rosenberg, D. (1994). Attitudes toward physical education: Their impact on how physical education teachers make sense of their work. *Journal of Teaching in Physical Education, 13*(4), 406-420.

Tannehill, D., & Zakrajsek. (1993). Student Attitudes Towards Physical Education: A Multicultural Study. *Journal of Teaching in Physical Education, 13*(1), 77-84.

Theodorakis, Y., & Goudas, M. (1997). Physical education interventions and attitude change. *International Journal of Physical Education, 34*(2), 65-69.

Thienes, G. (2008). *Trainingswissenschaft und Sportunterricht*. Berlin: Pro Business.

Thomas, J., Nelson, J., & Silverman, S. (2010). *Research Methods in Physical Activity* (6th ed.). Champaign, IL: Human Kinetics.

Thompson, P.D., Buchner, D., Pina, I.L., Balady, G.J., Williams, M.A., Marcus, B.H., Berra, K., Blair, S.N., Costa, F., Franklin, B., Flethcer, G.F., Gordon, N.F., Pate, R.R., Rodriguez, B.L., Yancey, A.K., & Wenger, N.K. (2003). Exercise and Physical Activity in the Prevention and Treatment of Atherosclerotic Cardiovascular Disease. A Statement from the Council on Clinical Cardiology (Subcommittee on Exercise, Rehabilitation, and Prevention) and the Council on Nutrition, Physical Activity, and Metabolism (Subcommittee on Physical Activity). *Arteriosclerosis, Thrombosis, and Vascular Biology. Journal of the American Heart Association, 23*, 1-8. Retrieved September 26, 2013 from: http://atvb.ahajournals.org/content/23/8/e42.full.pdf+html

Trafimov, D. (2007). Distinctions Pertaining to Fishbein and Ajzen's Theory of Reasoned Action. In I. Ajzen, D. Albarracin & R. Hornik (Eds.), *Prediction and Change of Health Behavior: Applying the Reasoned Action Approach* (pp. 23-42). Hillsdale, NJ: Erlbaum.

Triandis, H. (1971). *Attitude and attitude change*. New York, NY: Wiley.

U.S. Department of Health and Human Services (2000). *Healthy People 2010: Understanding and Improving Health* (2nd ed.). Washington, DC: U.S. Department of Health and Human Services.

U.S. Department of Health and Human Services (2008). *2008 Physical Activity Guidelines for Americans*. Washington, DC: U.S. Department of Health and Human Services.

Utsch, S. (2007). *Sprachwechsel im Exil: Die linguistische Metamorphose von Klaus Mann*. Weimar: Böhlau.

Valdez, L.A. (1997). *Attitudes toward Physical Education of Middle School Students and their Parents*. Doctoral Dissertation, University of Southern California, Los Angeles, CA.

Vallerand, R.J. (1997). Toward a hierarchical model of intrinsic and extrinsic motivation. In M.P. Zanna (Ed.), *Advances in experimental social psychology* (pp. 271-360). New York, NY: Academic Press.

Vallerand, R., & Losier, G. (1999). An integrative analysis of intrinsic and extrinsic motivation in sport. *Journal of Applied Sport Psychology, 11*, 142-169.

Vallerand, R.J., Pelletier, L.G., Blais, M.R., Briere, N.M., Senecal, C., & Vallieres, E.F. (1993). On the assessment of intrinsic, extrinsic, and amotivation in education: Evidence on the concurrent and construct validity of the academic motivation scale. *Educational and Psychological Measurement, 53*(1), 159-172.

Vanhees, L., Lefevre, J., Philippaerts, R., Martens, M., Huygens, W., Troosters, T., & Beunen, G. (2005). How to assess physical activity? How to assess physical fitness? *European Journal of Cardiovascular Prevention and Rehabilitation, 12*, 102-114.

Volkamer, M. (1997). Überlegungen zum Alltagsbewusstsein von Schülern und Lehrern im Sportunterricht. Vortragsfassung des Workshopbeitrages. In W. Miethling (Ed.), *Sportunterricht aus Schülersicht. Alltag, Alltagsbewusstsein und Handlungsorientierungen von Schülerinnen und Schülern im Sportunterricht* (pp. 49-60). Hamburg: Czwalina.

Wallhead, T.L., & Buckworth, J. (2004). The role of physical education in the promotion of youth physical activity. *Quest, 56*(3), 285-301.

Wang, C.K., & Biddle, S.J.H. (2001). Young people's motivational profiles in physical activity: A cluster analysis. *Journal of Sport and Exercise Psychology, 23*(1), 1-22.

Weiten, W., Dunn, D.S., & Hammer, E.Y. (2012). *Psychology Applied to Modern Life: Adjustments in the 21st Century*. Belmont, CA: Wadsworth.

Wolters, P. (2008). Koedukation im Sportunterricht – Zwischen Gleichheit und Differenz. In M. Hempel (Ed.), *Fachdidaktik und Geschlecht. Vechtaer Fachdidaktische Forschungen und Berichte – Heft 16*. Vechta.

Woolfolk, A. & Schönpflug, U. (2008). *Pädagogische Psychologie* (10th ed.). München: Pearson Deutschland.

World Health Organization (2000). *Obesity: Preventing and Managing the Global Epidemic*. Geneva: World Health Organization.

World Health Organization (WHO) (2012). *Global Recommendations of Physical Activity for Health*. Geneva: WHO Press. Retrieved September 26, 2013 from: http://whqlibdoc.who.int/publications/2010/9789241599979_eng.pdf

Wydra, G. (2001). Beliebtheit und Akzeptanz des Sportunterrichts. *Sportunterricht, 50*(3), 73-78.

Xu, F., & Liu, W. (2013). A Review of Middle School Students' Attitudes Toward Physical Activity. In L.E. Ciccomascolo & E. Crowley Sullivan (Eds.), *The Dimensions of Physical Education* (pp. 284-295). Burlington, MA: Jones & Bartlett Learning.

Zeng, H.Z., Hipscher, M., & Leung, R.W. (2011). Attitudes of High School Students toward Physical Education and their Sport Activity Preferences. *Journal of Social Sciences, 7*(4), 529-537.

Appendices

Appendix 1: Additional Tables not included in the main part

BMI Distribution of the Sample.

Statistiken

BMI

N	Gültig	322
	Fehlend	0
Mittelwert		20,3072
Median		19,9056
Modus		20,76
Standardabweichung		2,79497
Minimum		13,65
Maximum		33,70
Perzentile	25	18,5132
	50	19,9056
	75	21,6509

Übersicht über Hypothesentest

	Nullhypothese	Test	Sig.	Entscheidung
1	Die Verteilung von attitude_mean ist normal mit Mittelwert 3.56 und Standardabweichung 0.69.	Kolmogorov-Smirnov-Test einer Stichprobe	,090	Nullhypothese behalten.

Asymptotische Signifikanzen werden angezeigt. Das Signifikanzniveau ist .05.

SPSS Graph for the Kolmogorov-Smirnov-Test for Attitudes.

Regression for Physical Education Grade

Koeffizienten

Modell		Nicht standardisierte Koeffizienten		Standardisierte Koeffizienten	T	Sig.
		RegressionskoeffizientB	Standardfehler	Beta		
1	(Konstante)	4,211	,093		45,498	,000
	Sportnote	-,313	,041	-,391	-7,596	,000

Institutions used by Students

Häufigkeiten von $Inst_st

		Antworten		Prozent der Fälle
		N	Prozent	
Institution_st[a]	Sportverein	189	46,7%	58,7%
	Sport-/Fitnessstudio	52	12,8%	16,1%
	kultureller Verein, der Sport in seine Aktivitäten miteinbezieht	26	6,4%	8,1%
	Sonstige	50	12,3%	15,5%
	keine aktive Vereinsmitgliedschaft	74	18,3%	23,0%
	Ich weiß es nicht	14	3,5%	4,3%
Gesamt		405	100,0%	125,8%

Häufigkeiten von $Inst_pa

		Antworten		Prozent der Fälle
		N	Prozent	
Institution_pa[a]	Sportverein	95	24,5%	29,5%
	Sport-/Fintessstudio	92	23,8%	28,6%
	kultureller Verein, der Sport in seine Aktivitäten miteinbezieht	19	4,9%	5,9%
	Sonstige	39	10,1%	12,1%
	keine aktive Vereinsmitgliedschaft	102	26,4%	31,7%
	Ich weiß es nicht	40	10,3%	12,4%
Gesamt		387	100,0%	120,2%

Institutions used by Parents

Institutions used by Peers

Häufigkeiten von $Inst_fr

		Antworten		Prozent der Fälle
		N	Prozent	
Institution_fr[a]	Sportverein	231	49,9%	71,7%
	Sport-/Fitnessstudio	85	18,4%	26,4%
	kultureller Verein, der Sport in seine Aktivitäten miteinbezieht	33	7,1%	10,2%
	Sonstige	47	10,2%	14,6%
	keine aktive Mitgliedschaft	27	5,8%	8,4%
	Ich weiß es nicht	40	8,6%	12,4%
Gesamt		463	100,0%	143,8%

Reasons for Students

Häufigkeiten von $Rea_st

		Antworten		Prozent der Fälle
		N	Prozent	
Reasons_st[a]	zur Verbesserung der Gesundheit	176	11,4%	54,7%
	zur Verbesserung des Aussehens	155	10,0%	48,1%
	um dem Alter vorzubeugen	20	1,3%	6,2%
	um Spaß zu haben	242	15,6%	75,2%
	um zu entspannen	88	5,7%	27,3%
	um mit Freunden zusammen zu sein	173	11,2%	53,7%
	um neue Bekanntschaften zu machen	49	3,2%	15,2%
	um deine Leistungsfähigkeit zu steigern	161	10,4%	50,0%
	um dein Gewicht zu kontrollieren	138	8,9%	42,9%
	um dein Selbstwertgefühl zu steigern	66	4,3%	20,5%
	um neue Fertigkeiten zu erwerben	101	6,5%	31,4%
	um dich besser in die Gesellschaft zu integrieren	24	1,6%	7,5%
	um an Wettkämpfen teilzunehmen	81	5,2%	25,2%
	Sonstige	57	3,7%	17,7%
	ich weiß es nicht	16	1,0%	5,0%
Gesamt		1547	100,0%	480,4%

Prevention of Students

Häufigkeiten von $Prev_st

		Antworten		Prozent der Fälle
		N	Prozent	
Prevention_st[a]	ich habe keine Zeit	235	45,4%	73,0%
	es ist zu teuer	31	6,0%	9,6%
	ich habe keine Lust	84	16,2%	26,1%
	ich mag keine Wettkämpfe	37	7,1%	11,5%
	eine Verletzung/Krankheit	39	7,5%	12,1%
	keiner/keine meiner Freunde/Freundinnen will mit mir Sport treiben	24	4,6%	7,5%
	schlechte infrastrukturelle Möglichkeiten	11	2,1%	3,4%
	Sonstige	30	5,8%	9,3%
	ich weiß es nicht	27	5,2%	8,4%
Gesamt		518	100,0%	160,9%

Reasons for Parents

Häufigkeiten von $Rea_pa

		Antworten		Prozent der Fälle
		N	Prozent	
Reasons_pa[a]	zur Verbesserung der Gesundheit	196	18,5%	60,9%
	zur Verbesserung des Aussehens	80	7,6%	24,8%
	um dem Alter vorzubeugen	116	11,0%	36,0%
	um Spaß zu haben	107	10,1%	33,2%
	um zu entspannen	102	9,6%	31,7%
	um mit Freunden zusammen zu sein	50	4,7%	15,5%
	um neue Bekanntschaften zu machen	23	2,2%	7,1%
	um ihre Leistungsfähigkeit zu steigern	58	5,5%	18,0%
	um ihr Gewicht zu kontrollieren	131	12,4%	40,7%
	um ihr Selbstwertgefühl zu steigern	30	2,8%	9,3%
	um neue Fertigkeiten zu erwerben	26	2,5%	8,1%
	um sich besser in die Gesellschaft zu integrieren	14	1,3%	4,3%
	um an Wettkämpfen teilzunehmen	7	,7%	2,2%
	Sonstige	41	3,9%	12,7%
	ich weiß es nicht	76	7,2%	23,6%
Gesamt		1057	100,0%	328,3%

Prevention of Parents

Häufigkeiten von $Prev_pa

		Antworten		Prozent der Fälle
		N	Prozent	
Prevention_pa[a]	sie haben keine Zeit	233	46,9%	72,4%
	es ist zu teuer	21	4,2%	6,5%
	sie haben keine Lust	74	14,9%	23,0%
	sie mögen keine Wettkämpfe	20	4,0%	6,2%
	eine Verletzung/Krankheit	41	8,2%	12,7%
	keiner/keine ihrer Freunde/Freundinnen will mit ihnen Sport treiben	10	2,0%	3,1%
	schlechte infrastrukturelle Möglichkeiten	7	1,4%	2,2%
	Sonstige	32	6,4%	9,9%
	ich weiß es nicht	59	11,9%	18,3%
Gesamt		497	100,0%	154,3%

Reasons for Peers

Häufigkeiten von $Rea_fr

		Antworten		Prozent der Fälle
		N	Prozent	
Reasons_fr[a]	zur Verbesserung der Gesundheit	149	10,7%	46,3%
	zur Verbesserung des Aussehens	158	11,4%	49,1%
	um dem Alter vorzubeugen	29	2,1%	9,0%
	um Spaß zu haben	224	16,1%	69,6%
	um zu entspannen	77	5,5%	23,9%
	um mit Freunden zusammen zu sein	168	12,1%	52,2%
	um neue Bekanntschaften zu machen	55	4,0%	17,1%
	um ihre Leistungsfähigkeit zu steigern	106	7,6%	32,9%
	um ihr Gewicht zu kontrollieren	106	7,6%	32,9%
	um ihr Selbstwertgefühl zu steigern	55	4,0%	17,1%
	um neue Fertigkeiten zu erwerben	64	4,6%	19,9%
	um sich besser in die Gesellschaft zu integrieren	24	1,7%	7,5%
	um an Wettkämpfen teilzunehmen	75	5,4%	23,3%
	Sonstige	43	3,1%	13,4%
	ich weiß es nicht	56	4,0%	17,4%
Gesamt		1389	100,0%	431,4%

Preventions of Peers

Häufigkeiten von $Prev_fr

		Antworten		Prozent der Fälle
		N	Prozent	
Prevention_fr[a]	sie haben keine Zeit	198	34,9%	61,5%
	es ist zu teuer	38	6,7%	11,8%
	sie haben keine Lust	131	23,1%	40,7%
	sie mögen keine Wettkämpfe	28	4,9%	8,7%
	eine Verletzung/Krankheit	33	5,8%	10,2%
	keiner/keine ihrer Freunde/Freundinnen will mit ihnen Sport treiben	17	3,0%	5,3%
	schlechte infrastrukturelle Möglichkeiten	11	1,9%	3,4%
	Sonstige	32	5,6%	9,9%
	ich weiß es nicht	80	14,1%	24,8%
Gesamt		568	100,0%	176,4%

Appendix 2: Post-Statistical Tables as Guidelines for Discussion

Type of School and SES

Schulform * SES Kreuztabelle

Anzahl

		SES		Gesamt
		"good"	"bad"	
Schulform	Hauptschule/Werkrealschule	38	27	65
	Realschule	95	23	118
	Gymnasium	121	15	136
Gesamt		254	65	319

Exercise Frequency of Students and of Peers

Sporthäufigkeit der besten Freunde/Freundinnen * Sporthäufigkeit Kreuztabelle

Anzahl

		Sporthäufigkeit				Gesa
		5-7 mal pro Woche	3-4 mal pro Woche	1-2 mal pro Woche	weniger als 1 mal pro Woche	
Sporthäufigkeit der besten Freunde/Freundinnen	5-7 mal pro Woche	11	8	1	2	
	3-4 mal pro Woche	27	84	32	7	
	1-2 mal pro Woche	3	31	60	17	
	weniger als 1 mal pro Woche	2	6	14	17	
Gesamt		43	129	107	43	

Exercise Frequency and Physical Education Grade

Sportnote * Sporthäufigkeit Kreuztabelle

Anzahl

		Sporthäufigkeit				Gesamt
		5-7 mal pro Woche	3-4 mal pro Woche	1-2 mal pro Woche	weniger als 1 mal pro Woche	
Sportnote	1	18	42	14	3	77
	2	18	68	63	14	163
	3	6	19	26	20	71
	4	0	0	3	4	7
	6	1	0	1	2	4
Gesamt		43	129	107	43	322

Author Information

Rolf Kretschmann is currently a secondary school teacher in Germany, teaching Physical Education and Philosophy. He is a former Professor of Kinesiology and Director of Physical Education Teacher Education at the University of Texas, United States. His research focuses primarily on educational technology, health promotion and interventions, as well as evidence-based practices in physical education. At the time of this study's data collection and analysis, he was appointed "Wissenschaftlicher Mitarbeiter (Assistentur)" (equivalent to Assistant Professor) of Physical Education and Sport Pedagogy at the University of Stuttgart, Germany. Contact e-mail address: kretschmann.rolf@gmail.com

Daniel Wrobel is currently a secondary school teacher in Germany, teaching Physical Education and English. At the time this study was conducted, he was a graduate student at the Department of Sport and Exercise Science of the University of Stuttgart, Germany, and grew interest in the topic of school students' attitudes toward physical education preparing for his final thesis. Contact e-mail address: dee-wee@web.de